D1033133

CONTROL THEORY IN BIOLOGY AND EXPERIMENTAL PSYCHOLOGY

CONTROL THEORY IN BIOLOGY AND EXPERIMENTAL PSYCHOLOGY

F. M. Toates, M.Sc., D.Phil.
Institute of Physiology, Odense University, Denmark

HUTCHINSON EDUCATIONAL

Hutchinson Educational Ltd
3 Fitzroy Square, London W1

London Melbourne Sydney Auckland
Wellington Johannesburg Cape Town
and agencies throughout the world

First published 1975

Set in Monotype Times

Printed in Great Britain by
William Clowes & Sons, Limited
London, Beccles and Colchester

ISBN 0 09 119660 4

CONTENTS

PREFACE

Control theory and computer simulation techniques are playing an ever increasing role in the sciences of biology and experimental psychology. The physiologist Guyton (1971) has recently revised his standard textbook of physiology, and devotes half of the first chapter to control theory. The experimental psychologist Sutherland (1972) believes that signal detection theory and control theory are the only two branches of conventional mathematics to have proved to be useful in psychology. It might be added that this is true despite numerous attempts to involve mathematics in psychology. In some branches of biology – for example, population, growth, and insect behaviour studies – simulation and systems methods are central to much of the discussion.

As highly relevant as these techniques are, the biologist, physiologist or psychologist has at the time of writing nowhere to turn when he or she wants an easily readable account of them. The object of writing this book was to provide just such an introductory account. I attempt to explain control theory and associated simulation techniques to the reader who is not necessarily well equipped mathematically, but at the same time I hope that it will appeal to the mathematically sophisticated who want to explore biological control systems. Perhaps, firstly, I have in mind the biology or psychology student who wishes to understand control theory and who has done mathematics but has forgotten much of it, or who never really deeply and fully understood it first time. In this sense it is an introductory text. At the same time I present a view of what I think systems physiology and physiological psychology are all about, and I hope that this will provide a source of discussion within physiology and psychology. I also present the relatively 'sophisticated' concepts of frequency response, stability and the complex plane.

The fact that the book is directed to such a wide spectrum of readers has inevitably presented enormous problems when deciding at what level to base it. I hope that the reader bears with me on this point. I have tried to write it in such a way that it is understandable to almost any person motivated to read it. Unfortunately for the reader with a sound mathematical background, this has meant simple explanations that may sound somewhat patronising. For this reason the mathematically competent reader can jump over at least the first parts of Chapter 2. The reader who is less competent mathematically may have a little trouble with Chapter 7, but I hope not.

Unfortunately, in the past some texts and most articles that have attempted to relate control theory to the biologist have seemed to the beginner to do little more than exhibit the intellectual muscles of the author. The bewildered reader is taken through *n*th-order differential equations, *s* and *z* planes, eigenfunctions and unrealisable prediction operators before he or she can mathematically crawl. At the end of the discussion the reader is frustrated with the mathematics and has lost sight of the original biological problem that the mathematics should be illuminating. My intention here has been to use simple concepts to bring alive what is going on in biological control systems. The reader is brought to a position where he or she can read and understand most of the published articles in this area.

I make no apologies for using a simple language. In my opinion it is crucial that the reader should be able to form vivid impressions of what is going on in a system. This cannot be achieved merely by a statement such as 'it may be seen that' followed by pages of equations that the student then faithfully reproduces at some subsequent date.

A bucket of water with a hole in the bottom means more to most people than a differential equation. This is not to underestimate the importance of differential equations. Quite the contrary, the bucket of water is an excellent means for understanding them. It merely means that so much is to be gained by working through simple examples as a step towards understanding control systems.

It must be emphasised that there is nothing inherently difficult or confusing in the concepts of feedback control theory. Indeed, the very essence of this approach, which is always well emphasised by its practitioners, is that it provides an explanatory tool for the biologist. The problem hinges upon presentation.

This book is primarily directed to the biologist and experimental psychologist. Perhaps, unfortunately, the examples chosen to illustrate control theory reflect the bias of the author towards physiology and experimental psychology. However, since the behaviour of systems is independent of the hardware with which they are constructed, the book should be relevant to any branch of biology where a systems approach is involved, and might also be appropriate for engineers, physicists or economists.

As far as control systems analysis is concerned, the book is a statement of facts, and yet it is also a statement of a particular point of view when it comes to emphasising various approaches. The viewpoint is that arrived at after five years at the Laboratory of Experimental Psychology, University of Sussex, working with Dr K. Oatley. During this time I was supported by the Science Research Council and The Medical Research Council.

Odense, Denmark F. M. TOATES

ACKNOWLEDGEMENTS

I am most grateful to my associates at Odense University for help in preparing this book, particularly to Søren Jacobsen for his critical comments. Dr K. Oatley, A. Hazlerigg and J. Cram of Sussex University were of immense help to me during the time I was writing the book, and I wish to thank them for their assistance.

I am also grateful to the following for permission to reproduce material: The Institute of Measurement and Control, Academic Press Inc. (London) Ltd., W. B. Saunders Co., The MIT Press, Macmillan Journals Ltd., The Institute of Electrical and Electronic Engineers, Inc., CBS Education International, The American Physiological Society, Brooks/Cole Publishing Company, The Journal of Physiology, The British Psychological Society, McGraw-Hill Book Company, Medical and Biological Engineering, and Society for Experimental Biology Symposia.

1 | INTRODUCTION

It must be relatively uncontroversial to start a discussion on the assumption that science attempts to gain an understanding of how things function. This involves rationalisation and simplification; often a variety of phenomena can be brought under a single explanation. From the understanding that is obtained thereby, an attempt can be made to extract laws that may then be used to take our knowledge further. For instance, it has always been widely observed that objects tend to fall towards the ground, but the velocity characteristics of this phenomenon were not satisfactorily formalised until Newton presented his laws of motion. These laws state that the force acting on a body equals its mass times its acceleration and that, therefore, a body remains at rest or travels at a constant velocity unless acted upon by a force. The falling body accelerates in the direction of the force, i.e. towards the centre of the earth, and the acceleration is proportional to the net force acting.

Newton gave a precise account of this phenomenon in terms of force, acceleration and the parameter of the system, mass. When we write $F = ma$ we are not explaining why a force accelerates a mass, rather we are characterising the phenomenon in terms of a stimulus (force), a response (acceleration) and the system that mediates the stimulus-response (mass).

Newton's laws give a formal theoretical basis for the dynamics of bodies. There is a precise correspondence between the law and the observed behaviour of bodies – indeed man was able to design spacecraft on the basis of the laws. The laws of motion also illustrate the inadequacy of man's unaided intuition; before Newton it was believed that a force must be acting on a body for it to move, rather than to accelerate. Where a theory does not fit the facts it should be an indication that the theory is either wrong, inadequate or irrelevant; before Newton the 'theories' clearly did not fit the facts.

The biologist has a certain amount in common with the physicist, but I believe that in the area we will discuss he has somewhat more in common with the engineer. An example from electrical engineering will serve to introduce the argument. Materials offer resistance to the flow of electric current; the size of the resistance depends upon the composition of the material. The physicist working in this area is interested in explaining resistance in terms of temperature and free electrons. However, the engineer need not be concerned with why a resistor offers resistance to electric current flow. It might be enough for him to know that the electric component that he is working with is a resistor having a value of 10 ohms. He is more interested in how he can use the established characteristics of components such as resistors in building complex electronic circuits. As far as he is concerned the resistor is a 'black box' that relates an electric current (the output) to an applied voltage (the input) by means of the equation: resistance = voltage/current.

Let us now return to mechanics, and consider the problem of a weight falling through a fluid medium, our object being to establish a mathematical description of the behaviour of the weight. We can either make purely empirical observations and base our equation upon these, or we can attempt to explain the performance in terms of basic physical principles and then compare the performance predicted theoretically with that actually observed. If we were to drop the weight several times we might find that it accelerates from zero and reaches a maximum velocity of approximately x km/s at time t_1, and therefore we could establish an equation or several approximate equations to describe its velocity characteristics. However, it might be more profitable to attempt to analyse the system in terms of the forces that we believe are acting. In fact we need three basic equations:

1. Newton's law states that: force = mass × acceleration.

2. The net force on the body is the force of gravity minus the frictional drag of the fluid.

3. Frictional drag is proportional to velocity.

Thus $$F_1 - F_2 = ma \tag{1a}$$
$$F_2 = Kv \tag{1b}$$

where K is a constant, v is velocity, F_1 is the force of gravity, F_2 is the frictional force, a is the acceleration, and m is the mass.

The performance of the mass could in all probability be accounted for in these terms. We must now ask ourselves what exactly it is that we have done. The answer might be as follows. We believe there to be basic equations that define the behaviour of the system. We have assumed the validity of these, and then we have compared the overall performance of the system to the predictions arising from the combination of the basic equations. If the actual performance departed from that predicted then this would indicate that the equations employed were insufficient and that possibly there is non-linear friction component present. Clearly, what we ought not to do is immediately to change our three equations if the predictions of these equations do not match the new system.

In constructing the model represented by equations (1a) and (1b) we have not tried to prove or disprove Newton's law. Neither have we presented an explanation of why a force accelerates a mass. Rather we have accepted and employed some fundamental assumptions, put them together, and obtained an explanation in terms of the combination. The assumptions on which the model is based are quite unambiguous, as are the predictions that it makes. The model embodies our understanding of the component operations that go into making up the real system. The basis of the philosophy advanced in this book is that the same tools employed for the analysis of physical and engineering systems may be extended to the domain of biology. In particular, this book is addressed to the question: 'Can we explain certain aspects of physiological regulation and homeostatic behaviour in terms of the interaction of the physiological components that we know make up the system?'

Though the engineer is building a complex system from simple components such as resistors, and the biologist is analysing a system already constructed, it is remarkable just how much in common the two approaches have. When the physiologist considers the circulatory system he is not concerned with the chemical composition of the red blood cells, but with their mechanical properties, such as viscosity. Similarly, he is prepared to accept the equations for fluid flow in terms of the size of blood vessels, these equations having been borrowed from engineering. The experimental psychologist studying Pavlovian conditioning is not primarily concerned with the properties of individual nerves, rather it is the possible mode of interconnection that is his first interest. Both the physiologist and the experimental psychologist are trying to understand the behaviour of systems in terms of the known properties of their subsystems. Each of these subsystems will in turn be further divisible into subsystems that will be another person's area of interest. In the case of Pavlovian conditioning, the researcher who is trying to establish the basis of neural conduction (the subsystem of the experimental psychologist) will accept the movement of ions in response to an electric gradient as being an established fact.

A model of a system is a theory of how a system functions but, as opposed to so many rather vague theories in biology, it is a theory in terms that are unambiguous. A theory should be an explanatory tool in that through the theory we attempt to understand the real system. We extract from our observations and measurements on the real system certain deductions that we build into a theory. If we can, for example, account for the clinical phenomenon of oedema in terms of measurable properties of the body fluids and their compartments, such as interstitial compliance and colloid osmotic pressure, by building these characteristics into a model that itself exhibits 'oedema', then we can justifiably claim to have produced a theory of the formation of oedema.

At first the flow of information is from the real system to the model, in that we build a model based upon the characteristics of the real system. We then examine the overall behaviour of the model and compare this with the behaviour of the real system. Temporary changes can then be made to parameters to discover which of them are crucial for the performance of the model, and presumably which are crucial to the real system. Quite unexpected explanations appear at this stage, and so we now have a two-way flow of information. Physiology would seem to consist essentially of explaining how things work, analogous in our terms to a flow of information from the model to the real system.

A comparison with physics and engineering shows that from observations and careful measurements laws are extracted and models constructed that account for the behaviour of a mechanical system in terms of force, acceleration, friction, elasticity, inertia, etc. In engineering, a model may take the form of an electronic computer simulation in which electronic components form the analogue of mechanical components. To give a particularly relevant example, one can take the case of the design engineer engaged in the problem of space travel. The expense and risk to human life involved in travelling to other planets means that the engineer must spend a lot of time working with models. Questions are asked, such as, suppose the gain of component x is increased and at the same time that of y decreased, what effect will this have on the speed of response? In this case the information flow is from the model to the real system.

Despite the differences, is it not possible to compare the task of design engineering with that of medicine? Both essentially ask the question: How can we change the performance of a system? Typically the medical problem could be: Is the effect of a combination of drug x and drug y more potent than the effect of each individual drug? If not, why not? We have in a model a precise tool for testing the effects of drugs, surgical manipulations, or electrolyte deficiencies. Not only can we make empirical observations at a safe distance from the patient, but in addition we have a precise theoretical basis for explaining the effect of the manipulation that we are performing.

In building models of biological systems we are not trying to get the correct output from our model for a particular input, as an end in itself. That is usually not too difficult to accomplish and is analogous to the mechanical engineer who watches a mass falling through a fluid and tries to fit the best curve he can think of to the data obtained. Rather, we are building into the model identifiable properties of the subsystems and seeing how the combination behaves. This is like the analyst who bases the model of a mass falling through fluid on the laws of physics. The need for a model of a biological system to exhibit the same structure as the real system is emphasised very vividly by MacKay (1956):

An automaton giving the most perfect imitation of observable human behaviour might prove to hold as much interest for the psychologist as a gramophone capable of reproducing human singing does for the laryngologist, and no more.

By now one of the aims of the book may have become clear, and that is to describe mathematical model building. By mathematical models I mean models that relate to identifiable aspects of the real system and embody its structure, or at the very least refer to meaningful terms. The primary aim of the book, however, as implied by the title, is to discuss control systems. These are a special class of system, but none the less all of the principles that have been described for the building of models are relevant to these systems. Having said that I must unfortunately report that the reader who is expecting to find in the text a representation of the body in the form of exact, neat 'wiring diagrams' will be disappointed. Much of the discussion of control systems will be qualitative. The aim will be to show the basic principles, with an idea towards a complete understanding in terms of quantitative equations; in some cases it will be shown that models based upon a physiological understanding have been able to explain aspects of behaviour.

Control systems have long been a part of the engineer's concern; of particular interest are a class known as negative-feedback control systems. In building a process plant for the production of a chemical having a particular composition, the engineer does not base the design on the assumption that everything will work exactly as predicted at all times. On the contrary, the design enables the system as far as possible to compensate automatically for disturbances. It also makes sure that unintended changes in the characteristics of the components of the system have a minimum effect on the end product. This means that the end product is continually observed and compared with some measure of what is required. The control action is made to be dependent upon this comparison.

The same characteristics are exhibited by biological systems. The system that controls body temperature compares actual body temperature with some norm or reference value and the action taken depends upon the outcome of this comparison. Disturbances to the system (for example, getting into a cold bath) are remarkably well compensated for. Since there is a parallel between engineering control systems and biological control systems, and since engineers have accumulated a wealth of knowledge on the behaviour of feedback controllers, it seems a worthy area for the exchange of ideas. Many biologists have recognised this, but others remain unconvinced. The disparity of opinions is reflected in comments by Oatley (1972):

> The application of control theory to biological systems is seen by some as being quite essential to their understanding, in something like the way that an ability to count is necessary for knowing whether one's change is correct. By others it is seen as yet one more fad whereby those with some faintly exotic expertise can rephrase what is known already in terms that serve to obscure it.

Perhaps the reason that so many biologists fall into the second category is that control systems theory seems so often to be irrelevant to the problems that they are confronting. Even if they suspect its relevance, on closer examination it appears to be an incomprehensible language. I believe that this is because control theorists working in biology have often failed to relate their work to the body of knowledge already available. One ought not to be surprised if the laryngologist is not interested in a gramophone, and one should be no less surprised if the biologist is not interested in a set of numbers that he cannot relate to terms with which he is already familiar. The ophthalmologist, for example, is unlikely to be impressed by pages of differential equations involving half of the Greek alphabet, but on the other hand a theory that attempts to explain a clinical phenomenon in terms of the construction of the eye-movement controller may be more successful. I am not arguing that differential equations are unimportant, on the contrary, but the message presented in this book is that the control theorist should keep his feet on the ground and remember that others with less expertise in mathematics ought to gain something from the results.

So often the control theorist is found secreted away in a department called Cybernetics, Bio-Engineering, Mathematical Biology or Systems Theory, and while this reflects the inevitable specialisation of present day science, it also means a certain amount of barrier-building in relation to the broad area of biology.

The view that is advanced in the present text is that control systems techniques, borrowed from engineering, are central to the process of biological explanation. It is probably not an exaggeration to say that they are as relevant as maps are to understanding geography. In biology the 'hardware' is of course quite different from that with which the engineer is familiar and, in addition, the complexity of the biological system can be intimidating. However, hardware is probably the least important factor in systems analysis, since the same equations may be employed to describe systems as different as are an electronic circuit and the renal excretion mechanism. Within engineering itself, electronic models are frequently constructed as analogues of mechanical systems, and the two essentially exhibit exactly the same behaviour. As to the second point, the complexity of biological systems provides a compelling reason for attempting to build formal mathematical models, however inadequate they may subsequently appear. The human brain is often quite incapable of appreciating the consequences of the theories that it is able to propose. It is all too easy to make mistakes in logic when proposing an explanation in words, and to pass over the mistake repeatedly. A computer model that embodies our assumptions will ruthlessly expose any weaknesses that are inherent in our theorising, on the other hand it will also reveal unexpected explanations. It is impossible to imagine how one could fully understand a system as complex as the body-fluid regulation system without the assistance of a computer model. The computer will present us with an unbiased account of the outcome of our assumptions, something that we might be incapable of doing even with the most honest intentions. If the computer makes a mistake it will usually be obvious. The digital computer might present us with an answer like £ where we expect to receive 6·3. It is perhaps somewhat less obvious in the case of the analogue computer.

If the impression has been given that the present state of biological control systems analysis is that of an inaccessible subject out of the mainstream of biological thinking, then this has been unintentional. It would certainly appear that many analyses have been left unrelated to the existing state of knowledge, but at the same time the influence of control theory in biology has been quite remarkable. The physiologist Milhorn (1966) has built up a most powerful case for systems physiology which concludes with the following remarks:

> ... no model of any physiological system has yet attained its final state. Therefore, one could, at random, pick a system to study (using control system concepts) and turn out original research of some kind or another.
>
> One virtually sees physiology in a new light. He begins to look at the entire system and to think about how it functions as an entity. The literature is full of information gaps which become evident upon using this approach. Thus, new experimental ideas suggest themselves. Once the model is completed, it may also suggest new experimental research (necessary for verification of predictions by the model). Perhaps the most important thing that comes from the application of modelling, and control theory in general, is the evolution of a new way of thinking about physiological systems.

After surveying the literature and examining conference proceedings, one is forced to agree with Milhorn's final remark. It seems to be almost impossible to discuss physiology without reference to control theory. What started as little more than a fascinating intellectual game has almost revolutionised the science of physiology. Homeostasis has of course long formed one of the fundamental phenomena of physiology, but it is relatively recent that physiologists have come to accept that a formal embodiment of the homeostatic concept can only be made in control theory terms.

Homeostasis also forms the basis of one of the areas of interaction between psychology and control theory. The homeostatic controllers of body water and energy involve behaviour as part of the system, and indeed the behaviour studied primarily by the psychologist interested in motivation. Thus there have been and continue to be attempts made to define motivated behaviour in control theory language. A model of the body fluid-drinking system is a most useful tool for understanding such topics as thirst satiety, osmotic thirst and haemorrhage-induced drinking; it enables behaviour patterns to be related to quantitative changes and to the recognisable physiological components of the system. In this way it represents a theory of motivation, but at the same time it is important to recognise its limitations. We should not be tempted to bend the facts in order to fit the theory, and we must acknowledge that such a model can probably say nothing about the excessive amounts of water that a rat drinks when food pellets are presented to it at a rate of about 1 per minute.

Of course we could always, with perfect honesty, make the computer model 'drink' in response to interval food schedules. All we need do is to specify in the programme when the pellets are arriving and build into our drinking circuits a term that says 'if a single pellet of food has just arrived allow a few seconds to elapse and then drink 0·3 cm^3 of water'. An accurate simulation of drinking could be produced in this way, but it would tell us nothing about the basis of drinking motivation (see Oatley (1973) for a further discussion of these points). The reason that such an addition to the model would tell us nothing is that we have observed a behaviour pattern and merely copied it. The operation performed by the computer in obtaining the desired result need not relate to the basis of the animal's behaviour. For something to have explanatory value a question must be posed in the first place, and there must be some doubt surrounding the outcome of the question. A reasonable question might be: Given a physiological understanding of the processes involved in digestion and energy utilisation as well as the hunger mechanism, can we then explain eating behaviour?

Physiological psychology is one fertile area for control theory, but it is probably in the area of human performance that control theory has had most influence in psychology. The original observations that man behaves as a negative feedback controller in performing movements brought a new word into vogue: cybernetics, the study of control and communication in the animal (Wiener, 1948). In addition, the emergence of the new applied science called ergonomics has provided the basis for developing the idea of man as a system, with the emphasis not on man as a self contained system but as a system interacting with the environment. Ergonomics is concerned with the study of man–machine and man–environment interactions, fortunately with a view towards fitting the machine or environment to the man rather than vice versa. Those who call themselves practitioners of ergonomics have a wide variety of interests ranging from designing comfortable shoes and furniture, and training olympic athletes, to determining illumination levels in offices. However, the area of ergonomics that concerns us here is that of man as the operator of a machine, for example, an aircraft. The aircraft designer is able to quantify reliably the performance characteristics of the resistors and capacitors, levers and wing flaps, etc., that form components of the control circuits – but then at the centre of the system and forming an integral part of it there is man. The question is can we or can we not extend our mathematical model-building to include man's input–output characteristics? The view of ergonomics is that in some respects we can quantify the expected behaviour and performance limitations of the human operator. We must at the same time acknowledge that man is an adaptive, learning, and at times quite unpredictable component.

Ergonomics sometimes appears to have a sinister flavour about it. It is sometimes argued, in the context of industrial production, that to consider man as a component of a system can only serve to advance the nightmare of a computer controlled society where man has a status little better than the machine with which he works. Such arguments are probably quite justified up to a point, but we should also consider the positive side. Designers employing the ergonomic philosophy have been responsible for many innovations that have improved working conditions, for example, the design of cranes that do not involve the operator making uncomfortable bodily contortions in order to see the load.

The quantification of reaction time as a component within the human controller means that tracking tasks may be designed so as not to pose unreasonable demands on the operator. The human 'computer' has its own merits, for example, its ability to make intelligent decisions in response to the unexpected. However, we must also recognise that it has its limitations; the slowness with which it performs calculations, such as addition and differentiation, when compared to an electronic computer is an example. An understanding of these characteristics means that in a tracking task, where the human operator forms part of the system, certain operations can be performed by the machine while others are made the responsibility of the operator.

At times one can not help feeling that machine-like explanations of behaviour are somewhat depersonalising. But one must remind oneself that psychology is in dire need of new explanatory tools, many older methods having failed to give a convincing explanation of behaviour. Apart from control theory, the new tools appearing in psychology include artificial intelligence involving visual and auditory pattern recognition. In common with control theory, these involve modelling, in another domain, the kind of logical processes that the brain might conceivably employ.

The analogy between the heart and a sump-pump has proved to be useful in understanding the circulation of blood and at the same time has not resulted in the raising of protests. One might conclude, therefore, that the benefits neurophysiology and psychology can derive from borrowing the techniques and principles of engineering outweigh the disadvantages.

An area of interaction between the engineering and biological sciences, that is central to the present discussion, is that of designing artificial aids. In the construction of an artificial limb the designer needs to take not only the anatomy of the body into account but also the fact that a normal limb would provide the brain with information about position and muscular tension. It may be possible that the subject can learn to rely upon visual cues, but whatever is the case the bio-engineer should be familiar with the problem as a whole. To give another example, in the case of building an automatic range-finder for the blind, one of the first tasks that the research team set themselves was to build a model of the accommodation control system of the normal eye (Crane, 1966).

In advancing the argument that the control techniques used in engineering are also the relevant tools for the biologist, one is in a sense merely building upon an already established area of contact between these sciences. Equations have been applied to a number of physiological components, for example the relationship between blood flow through a vessel Q, resistance R and pressure difference across the vessel ΔP is given by $Q = \Delta P/R$. In constructing a control model of the circulation, one would employ equations of this kind as the components and then combine them. To demonstrate the essence of the systems approach, one can consider a very simple example of a lever, as shown in Figure 1(a).

If displacements x are given to the lever at the left, and measurements of the displacement in terms of distance around the circumference are made, the corresponding displacements y at the other end can be observed. From simple mathematics it is easy to calculate that the relationship between x and y is given by $y = Kx$, where K is a constant. The calculations would show that K has the value R_2/R_1, and the equation $y = (R_2/R_1)x$ has completely characterised the performance of the lever. Figure 1(b) shows a block diagram of the system. The input is x,

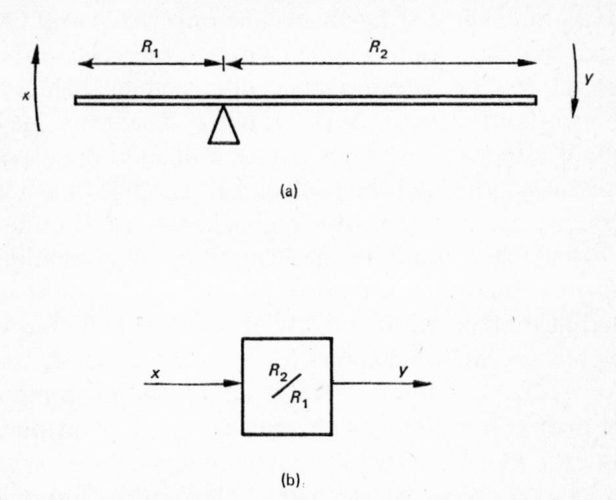

Figure 1 A model of a physical system. (a) The lever has displacements x applied to it and the displacement y is measured. (b) The block diagram characterises the performance of the lever. The displacement x is multiplied by the gain R_2/R_1 in order to give the displacement y.

the output y and the operator that relates input to output is the constant R_2/R_1. The block diagram style is attractive since it shows the signals or variables of the system as arrows, and the operators that characterise the system appear as the contents of the block. The performance characteristic of the lever is completely defined by the operator R_2/R_1. The block diagram approach provides a quite unambiguous language for describing systems, and ideally one would like to be able to characterise biological systems in these neat terms. This can be done to a very limited extent and Chapter 4 describes the construction of mathematical models.

Chapter 5 describes a special class of system: the negative feedback controller. The discussion is largely qualitative, reflecting the fact that many of the systems cannot be quantified. However, the principles of control theory are of first importance even if the nature of the system precludes a rigorous analysis at this stage. In Chapter 6, the mathematics of control systems is considered and it is shown how one can predict the performance of a controller on the basis of the components from which it is formed.

In Chapter 4, I deliberately speak first about systems without any reference to negative feedback, and only in the next Chapter introduce feedback control systems as a special class of system that are made up of the components discussed in Chapter 4. The main reason for this order of presentation is that one may wish to produce a model in systems notation of, say, the passage of sodium through the intestine wall, and at the same time not have the slightest interest in negative-feedback systems. It is therefore useful to be able to think of mathematical model-building in a context other than that of control theory. It is hoped that after considering mathematical models at some length, the reader will appreciate, when Chapters 5 and 6 are reached, that the negative-feedback system is made up of simple components that by their connections give rise to complex behaviour.

Chapter 7 describes frequency response methods, where in several cases a precise mathematical characterisation of a system has been possible. In Chapter 8 behavioural aspects of homeostasis are discussed in the context of a digital computer model of the thirst-body fluid control system.

In selecting examples to illustrate systems techniques, I hope that some kind of unity between physiology and psychology will appear and that systems techniques can serve to confirm this unity. For example, the homeostatic systems of temperature, body fluids, energy and sodium balance all involve behaviour as part of the control loop. Sexual behaviour, a non-homeostatic form of behaviour, is dependent upon hormonal factors which presumably may be defined by differential equations. Physiology, psychology and control theory each appear to have a part to play in understanding these problems.

However, before these subjects are presented, the essential mathematics is reviewed in Chapters 2 and 3. An attempt is made to produce a dialogue between the examples in mathematics and the examples of physical systems that are presented, so that at no time will a sudden discontinuity appear.

2 | BASIC MATHEMATICS

Some Rules and Definitions

The following rules of mathematics will assist the reader in understanding material presented further on in this book.

1. The equation:

$$x = 2a+2b$$

may be re-written as

$$x = 2(a+b)$$

in other words, a common term may be placed outside of brackets.

2. The rule for manipulating equations is that what we do to one side so we must do to the other.

Thus if

$$x = 2yz$$

dividing both sides by 2 gives

$$x/2 = (2yz)/2$$

$$x/2 = yz$$

3. If $\quad x = y+z$

then dividing both sides by 2 gives

$$x/2 = (y+z)/2$$

and since $(y+z)/2$ is the same as $y/2+z/2$ we have

$$x/2 = y/2+z/2$$

4. If $\quad x = yz$

to obtain z we divide both sides by y in order to isolate z

$$x/y = (yz)/y$$
$$x/y = z$$

5. If $\quad x = y+z$

to obtain y we subtract z from each side in order to leave y on its own

$$x-z = y+z-z$$
$$x-z = y$$

If $\quad x = y-z$

to obtain y we must add z to each side

$$x+z = y-z+z$$
$$x+z = y$$

6. One sometimes encounters equations of the form

$$x = (y/2+z/2)/(w/2)$$

This may be simplified by multiplying both top and bottom of the right-hand side by 2 which of course leaves its value unchanged. Thus

$$x = (y+z)/w$$

7. Pythagoras' theorem states that for a right-angled triangle (as shown in Figure 2) the square of the side opposite to the

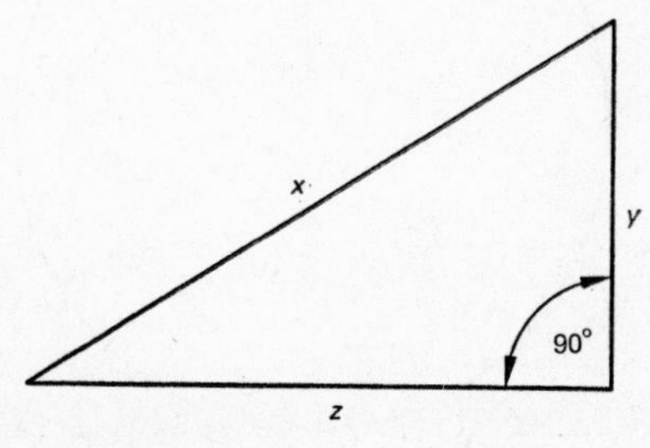

Figure 2 For a right angled triangle $x^2 = y^2+z^2$

right angle is equal to the sum of the squares on the other two sides, i.e.

$$x^2 = y^2 + z^2$$
$$x = (y^2 + x^2)^{1/2}$$

8. A circle is divided into 360 degrees, and Figure 3 shows the main landmarks around a circle, i.e. $\frac{1}{4}$(90°), $\frac{1}{2}$(180°) and $\frac{3}{4}$(270°) of a cycle measured in an anti-clockwise direction.

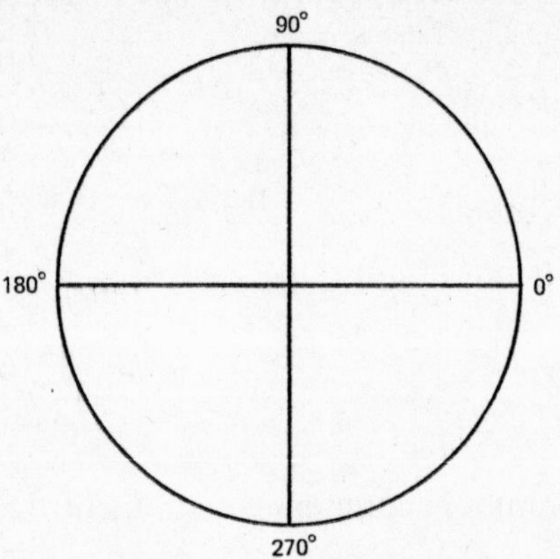

Figure 3   A circle may be divided into four equal parts each of which is 90°.

9. A radian is the angle (θ) subtended at the centre of a circle by an arc whose length is equal to the radius of the circle, as illustrated in Figure 4. Since the length of the circumference of a circle is given by $2\pi r$, there are 2π radians in 360°.

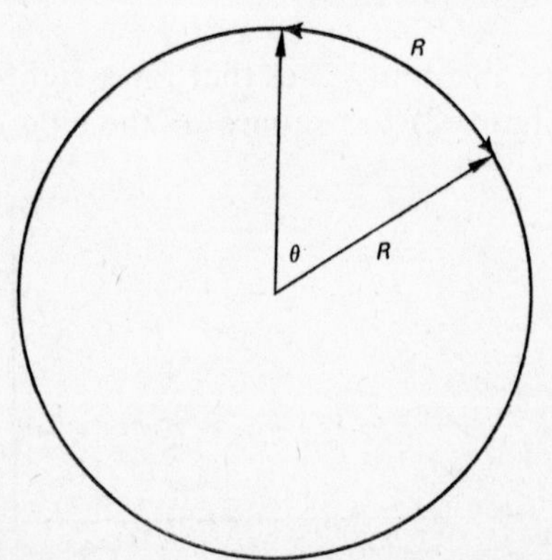

Figure 4   The radian is the angle subtended by the arc of length R.

Powers and Logarithms

Consider the case of 2^2 being multiplied by 2^3

$$2^2 \times 2^3 = (2 \times 2) \times (2 \times 2 \times 2)$$
$$= 2^5$$

this is the same thing as 2^{2+3}.

It is a general rule that when we multiply like terms that are raised to various powers then we can simply add the powers, i.e.

$$x^a \times x^b = x^{a+b}$$

From this we may also deduce the value of x raised to the power of zero.

$$x^a \times x^0 = x^{a+0}$$
$$= x^a$$
$$x^0 = x^a/x^a$$
$$x^0 = 1$$

Anything raised to the power zero is 1.

$$x^a \times x^{-a} = x^{a-a}$$
$$= x^0$$
$$= 1$$

Therefore $\quad x^{-a} = 1/x^a$

The rule here is that something raised to a negative power is the same as that term appearing in the denominator but with a positive sign.

The rule that $x^a \times x^b = x^{a+b}$ forms the basis of the mathematical tool known as the logarithm. Let us take the example of logarithms to the base 10. Given that $10^1 = 10$ we say that 1 is the logarithm of 10 to the base 10, in other words we must raise the base 10 by the power 1 to give the number 10. The logarithm of 100 to the base 10 is 2, i.e. $10^2 = 100$, and 3 is the logarithm of 1000 to the base 10.

If we know the logarithm of each number to a particular base then we can simply add the logarithms in order to multiply numbers. This gives us a logarithm as an answer and we can then locate which number corresponds to this logarithm in order to find the answer.

For example, let us find what is the product of 10 and 100.

$$\log_{10} 10 = 1, \quad \log_{10} 100 = 2$$

$$\log_{10}(10 \times 100) = 2+1 = 3$$

The number whose logarithm is 3 is 1000 (we say the anti-logarithm of 3 is 1000). The solution is that $10 \times 100 = 1000$.

Obviously, in practice we would not use logarithms to solve this problem, but the example serves to illustrate the essential principle involved. There are sets of logarithms available so that we may find the logarithm of 46·9 almost as readily as that of 10. It is when handling numbers of this kind that the technique is most useful.

The Sinewave

Consider the radius r of the circle shown in Figure 5(a) to rotate about the centre at a constant speed. Distance x represents the projection of the radius onto the vertical axis. If we measure the angle θ that the radius makes with the horizontal and plot x against θ, we will obtain the function shown in Figure 5(b). This is known as a sinewave. It starts from zero, reaches a maximum when the radius is vertical, falls to zero again and then takes a negative value when the radius goes below the horizontal. If we double the speed of rotation of the radius then we double the frequency (i.e. the repetition rate) of the sinewave.

When the radius has rotated through angle θ then the projection is x and, therefore, the height of the sinewave is x. We may therefore give the sinewave the units of angle corresponding to the angles of the circle that generated the sinewave. Thus, angle θ measures both the angle that the radius makes and the position on the sinewave. An angle of 90° means that the radius has moved through a quarter of a cycle and that the projection x has the value r, which is the maximum it can have. Corresponding to this, the angle 90° on the sinewave is the point of the maximum on the sinewave. 360° represents one complete rotation of the radius and also one complete cycle of the sinewave.

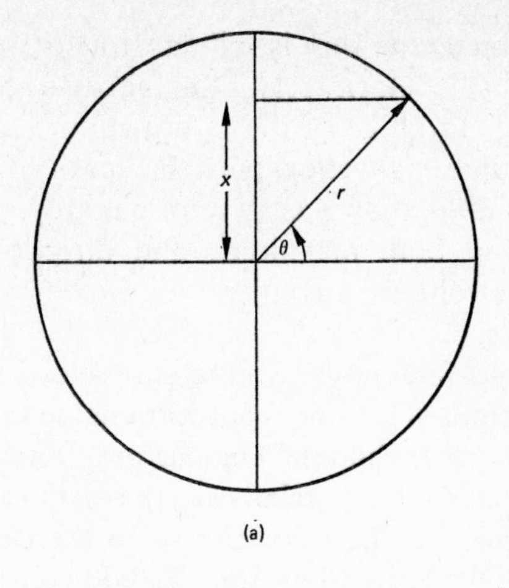

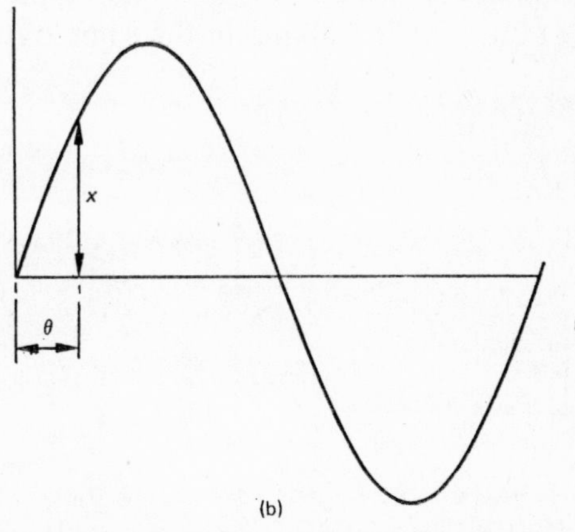

Figure 5 *A sinewave may be generated from the rotation of a radius.*

Vectors and Scalars

If we add 3 cm^3 of water to 2 cm^3 of water then naturally we obtain 5 cm^3 of water. Given that we are using the cm^3 as our unit of measurement, the numbers 3, 2 and 5 contribute all of the necessary information. The problem seems trivial, so let us move to another example. A ship is sailing at 10 km/h, and a man is walking on the deck at 5 km/h. What is the man's velocity relative to the earth? Here we cannot simply add 5 and 10 to get 15, unless of course we know that the man is walking in the same direction as the ship is sailing. If he is walking in the opposite direction his velocity relative to the earth is only 5 km/h. In this case then, to be given the information 5 and 10 is insufficient, in addition direction must be specified. Velocity is an example of what is known as a vector quantity, that is to say a quantity having both magnitude and direction. A scalar quantity, such as volume or temperature, is completely defined by magnitude alone.

Figure 6 demonstrates how we represent vectors. A line shows the vector, the length of the line is proportional to the magnitude of the vector quantity while the angular position represents its direction. The man's velocity relative to the earth is given by the sum of the velocity of the ship relative to the earth plus the velocity of the man relative to the ship. To add vectors we simply add their components in the horizontal and vertical directions. In Figure 6(a) the man is walking in the same direction as the

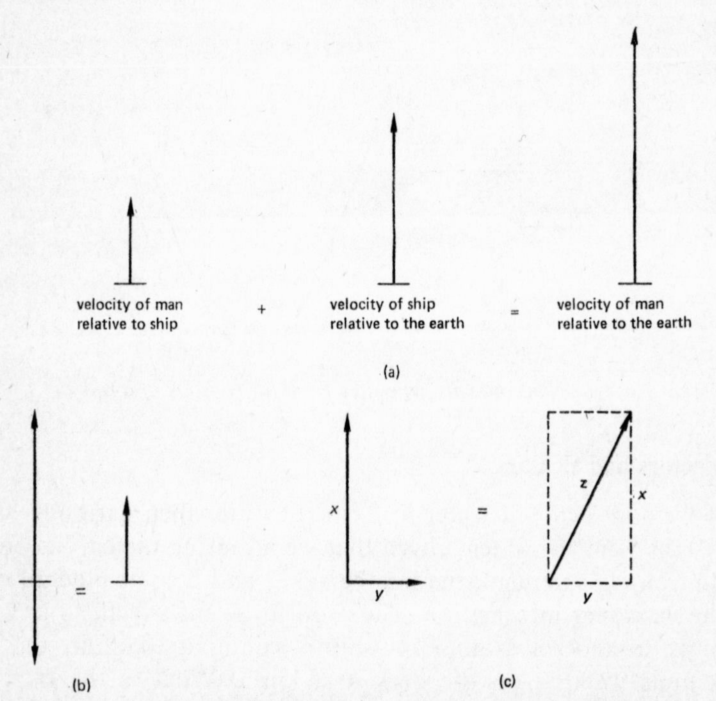

Figure 6 Addition of vectors. (a) Two vectors have the same direction, (b) they have opposite directions, (c) they act at right angles to each other.

ship, in (b) against the direction of motion, and in (c) at right angles to the direction of motion. In (c), the vertical component of the vector representing the ship is x and its horizontal component is zero, while the vertical component of the vector representing the man's velocity is zero and its horizontal component is y. The sum of these two vectors is a vector having the vertical component $x+0 = x$ and a horizontal component $y+0 = y$, i.e. vector **z**. No matter how many vectors we add, the sum is given by the sum of components in the two directions.

Another example of a vector quantity is force. It is necessary to specify the direction of two forces as well as their magnitude in order to determine the net force. If a force of 1 kg plus a force of 3 kg act on a mass the net force is 4 kg if they have the same direction, but 2 kg if they have opposite directions. The magnitude is $(1+9)^{1/2}$ kg if they are at right angles.

Complex Numbers

What is meant by a complex number may be understood by reference to Figure 7. The vector of magnitude 1 is drawn lying

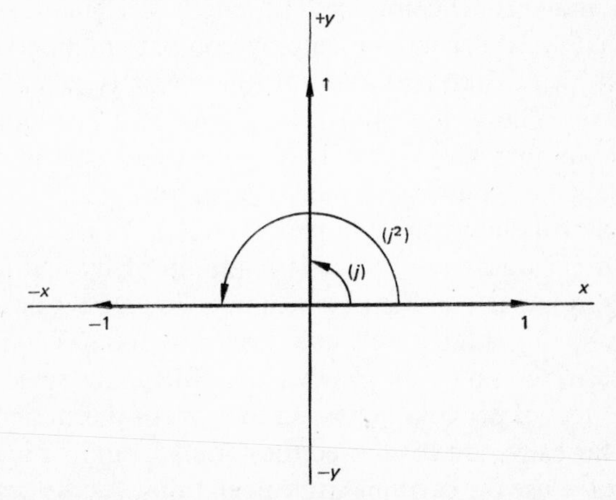

Figure 7 The complex plane. Real numbers are plotted along the x-axis while imaginary numbers are plotted on the y-axis.

along the x-axis in a positive direction. The vector of magnitude -1 is on the same axis but in a negative direction. In order to convert the positive vector into the negative vector it is necessary to multiply the vector 1 by -1. This has the effect of rotating the vector through an angle of 180°. Let us consider a new quantity which, when we multiply the vector 1 by it, has the effect of rotating the vector through only 90° so that it becomes positioned on the positive y-axis. If we again multiply the vector by this term, and it rotates it through a further 90°, the vector will find itself on the negative x-axis. Thus the effect of multiplying by this term twice is the same as multiplying by -1. If we call this term j, we can say that $j \times j = -1, j^2 = -1$ and $j = (-1)^{1/2}$.

The term j is used by engineers (i is usually used by mathematicians, but it means the same thing) to denote an operation which rotates another vector through 90° anti-clockwise. Terms involving j are known as complex numbers. These are vector quantities, and the use of the operator j provides a mathematical tool for carrying out operations on vectors.

One may specify a vector by giving its length and the angle that it makes with a reference direction. For example, the vector shown in Figure 8(a) is of length 5 and makes an angle θ with the horizontal. We could alternatively separate the vector into its components, and say that it has a horizontal component of length 4 and vertical component of length 3. The latter is essentially what we do when we employ complex numbers. The x-axis is known as the real axis and the y-axis is called the imaginary axis. The vector shown in Figure 8(a) is denoted by a complex number $4+j3$, that is to say its real part is 4 and its component on the imaginary or vertical axis is 3.

Figure 8(b) shows the complex numbers $2-j2$, $-2-j2$ and $0+j1$. In each case the vector is made up of its real and imaginary components, either of which may be positive or negative. The number $0+j1$ is a purely imaginary one and has no real part.

The term 'complex' is somewhat unfortunate since it often has the effect of provoking fear on the part of the student. However, as far as we are concerned the complex number is merely a tool that we use for carrying out calculations. As the example of a man walking on the deck of a ship showed, we need a tool that can account for the direction component as well as the magnitude component, and this is provided by the complex number. We break the vector up into its components in each direction, which conveys the same information as giving a distance and an angle.

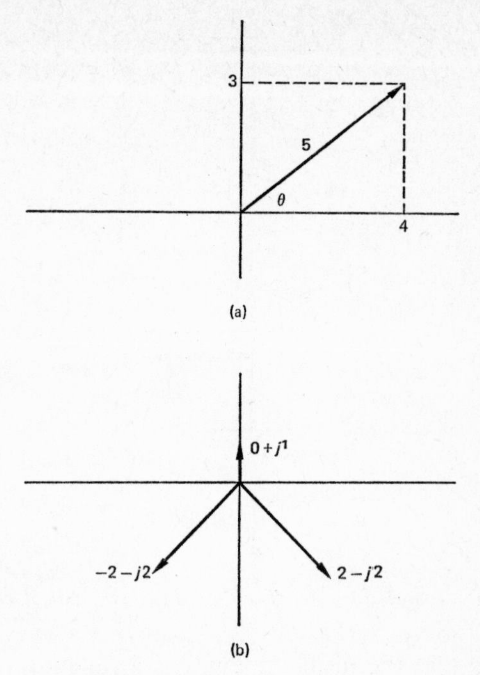

Figure 8 Complex numbers. (*a*) *The vector may be specified either in terms of its length* (5) *and its angle* (θ), *or in terms of its components in the horizontal or real direction* (4) *and the vertical or imaginary direction* (3). (*b*) *The complex numbers* $2-j2$, $-2-j2$ *and* $0+j1$.

A Function of a Variable

We will consider a variable y that is dependent upon another variable x such that for any value of x there is a corresponding value of y. We would express this in the following way

$$y = f(x)$$

where f means 'function of'. Figure 9 shows a spring and pointer with a mass attached that gives a force F_1 due to gravity. The dotted line indicates the position of the pointer when the force F_1 is zero, therefore distance x is the displacement caused by force F_1. We would write

$$x = f\,(\text{force})$$

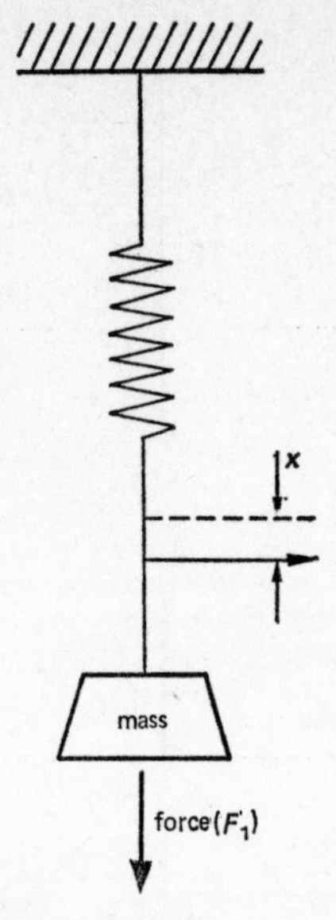

Figure 9 A force–spring system. The force F_1 causes a displacement x.

which reads that the displacement x is a function of the force that is applied. To give another example, the area of a square A is a function of the length of its side l, i.e.

$$A = f(l)$$

and to be specific

$$A = l^2$$

For the moment we will concentrate upon functions of one particular variable: time. Figure 10 shows a graph of the distance y that a car has travelled from its starting point as function of the time t measured from the point of starting. Since the function starts from the origin and is a straight line we are able to specify the relationship as

$$y = Kt$$

where K is a constant (K is usually chosen as the letter to represent a constant).

What we understand by this equation is that in order to obtain the distance travelled at any time t we multiply t by K and we

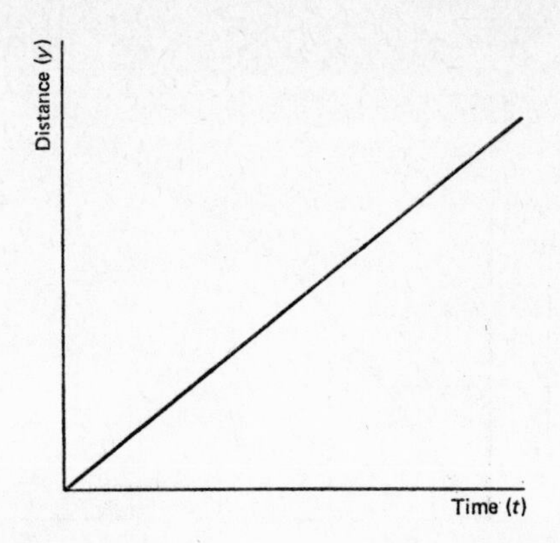

Figure 10 *The distance y that a car travels shown as a linear function of time t.*

obtain distance y. Sometimes we would see the equation written as $y(t) = Kt$ which reads that y as a function of time is given by K times t. If in one hour the car travels 50 kilometres we can specify K and therefore write

$$y = 50t$$

where t is the time in hours and y is the distance in kilometres.

DIFFERENTIATION

A car moves at a uniform speed and in 1 hour it has travelled 50 kilometres. What is the speed of the car? Of course it is 50 km/h, and the reader is probably wondering why the question is even posed. Although the answer is intuitively obvious, what we do to obtain the answer is to divide a distance by time in order to get a rate of change. In other words, every one is capable of doing differentiation. Speed is defined as: Δdistance/Δtime (Δ means change in), and in this case is given by 50/1, i.e. 50 km/h. It is only necessary to stretch the imagination a little after considering this example in order to understand differentiation completely.

Figure 11(a) shows a slightly more realistic relationship between the distance a car has moved and the time from its departure. How do we obtain the speed of the car from this graph?

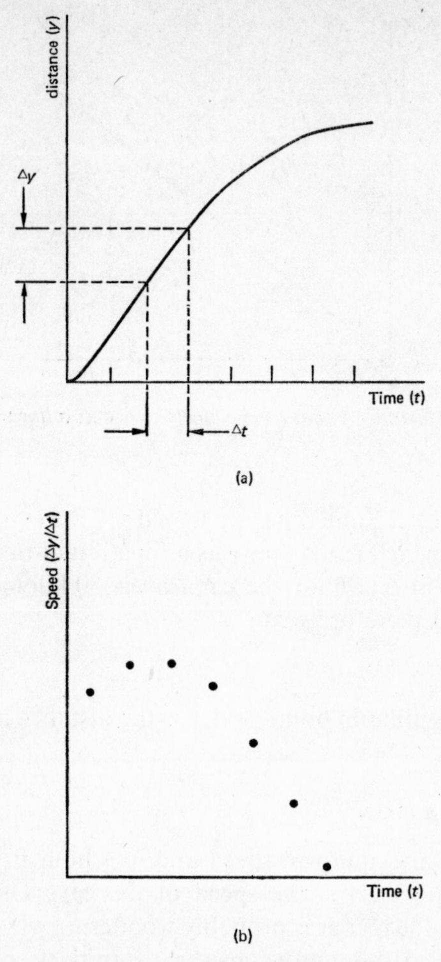

Figure 11 The relationship between the distance y that a car travels and time t, where speed is not a constant. The time axis is divided into short time intervals and for each of these the speed is calculated.

We have defined speed as a change in distance divided by the corresponding change in time. Figure 10 shows that the speed is constant and can be obtained by dividing an interval of distance by a corresponding interval of time. This is not the case for Figure 11(a), since the time axis has been divided into equal intervals and a different change in distance corresponds to each interval. What we can do is to consider the average speed over each interval. Over the interval of time Δt, the car travels Δy. If we divide Δy by Δt, then we obtain the mean speed over the interval Δt. Figure 11(b) shows a sketch of the result of calculating $\Delta y/\Delta t$ at each interval of time. If we were to make the time interval smaller and make more calculations then we would approximate the exact speed at any point in time. When Δt is made infinitesimally small, this is shown by replacing Δ with d, and the process of differentiation of y with respect to t is written as dy/dt, the rate of change of y with respect to t. Speed is the rate of change of distance with respect to time. By dividing $\Delta y/\Delta t$ we obtain the mean slope of the graph over the interval Δt. Another way of looking at the derivative is that it is the slope of the graph. Two new expressions can now be introduced: the derivative, and to differentiate. Speed is the derivative of distance with respect to time, and when we differentiate distance with respect to time we obtain speed.

Let us now suppose that the speed of a car increases at a constant rate, in other words the car has a constant acceleration. Acceleration is the rate of change of velocity with respect to time, it might be for example 2 km/h². Acceleration is therefore a derivative of a derivative, and if distance is y then acceleration is given by d^2y/dt^2.

Example 1

Sketch the derivative of the function shown in part (a) of Figure 12.

At first the function is linear, and so $df(t)/dt$ is constant. The function $f(t)$ then remains constant and so $df(t)/dt$ is zero. As Figure 12(b) shows, $df(t)/dt$ is a constant and then drops to zero.

Example 2

Make a rough sketch of the derivative of the function shown in Figure 13(a).

If we take a small interval of time Δt from time zero we find that $f(t)$ has increased to some finite value. Dividing this change by Δt we obtain value A for the derivative and this appears on the graph of $df(t)/dt$ against t. For a further small change in

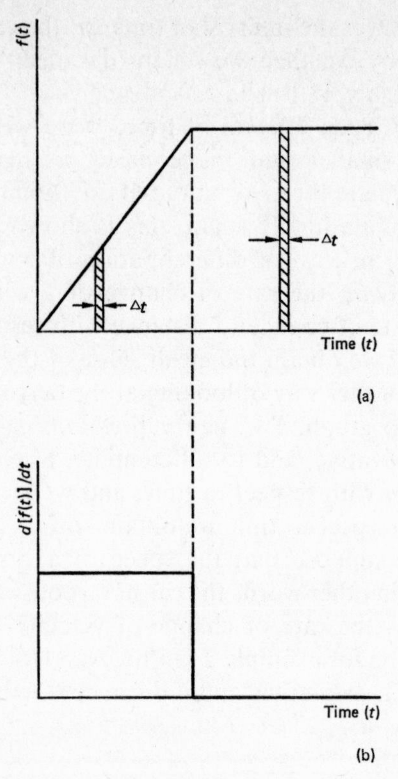

Figure 12 *See Example 1 for description.*

time we find that $f(t)$ has changed slightly less than the value A, and so the derivative takes a lower value. As we approach the peak of $f(t)$ the rate of change of this function with respect to time becomes smaller and smaller, until at the summit, $df(t)/dt$ is zero. After this point an increase in time is accompanied by a decrease in $f(t)$ and so $df(t)/dt$ takes a negative value.

INTEGRATION

This section can be initiated with a very simple question. If water flows into a container for 1 minute at a rate of 3 litres per minute, how much water does the container hold? We obtain the answer of 3 litres by multiplying time and flow, and in so doing have integrated flow with respect to time in order to

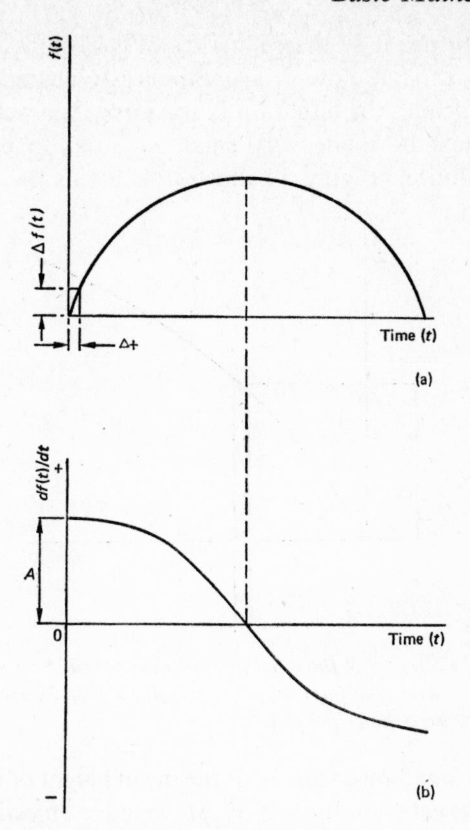

Figure 13 *See Example 2 for description.*

obtain volume. There will of course only be 3 litres in the container if it is empty at time zero. If a volume is already present then this must be added to the product of flow and time so as to obtain total volume. The quantity present at time zero is known as the initial condition.

Suppose the flow is 3 litres per minute for 2 minutes, 6 litres per minute for 1 minute and then finally $\frac{1}{2}$ litre per minute for 6 minutes. The initial condition of volume in the container is 10 litres. What is the volume at the end of 9 minutes?

$$\text{volume} = (3\times 2)+(6\times 1)+(\tfrac{1}{2}\times 6)+10$$

Obviously we multiply flow by the interval over which the flow occurs for each particular interval, and then add to this the initial condition. In this case the task is easy, since flow changes in a measurable way. If flow is a continuously changing quantity then, in principle, the situation is the same; however, our time intervals must be made very small in order to approach an accurate solution. Figure 14 illustrates this as the time axis is

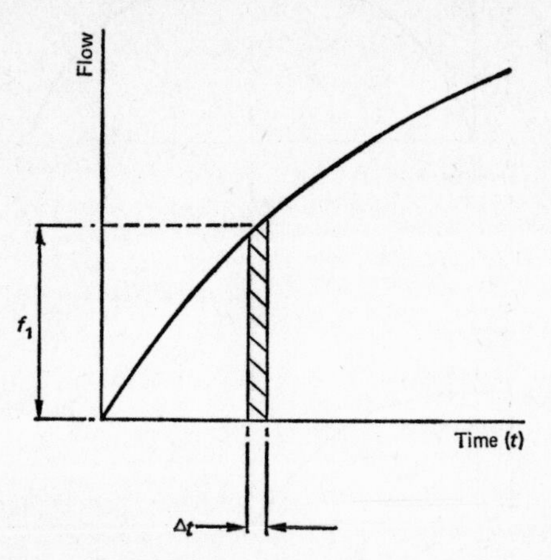

Figure 14 The integral of the function over the interval Δt is given by Δt multiplied by f_1. The total integral is the sum of all such components and is the same as the area under the graph.

divided into small intervals Δt. If the mean height of the function over the interval is multiplied by Δt, we have an estimate of the integral over the time interval Δt. Adding all such slices or sub-integrals together then gives us the total integral, the integral being nothing more than the area under the graph. When Δt is made infinitesimally small then the sum of all such areas accurately gives the area under the graph. Δt is replaced by dt in this case, and the integral of a function of time $f(t)$ over the period of time of zero to t is written as

$$\int_0^t f(t)\,dt$$

Volume is the integral of flow with respect to time, and distance is the integral of speed with respect to time. From these examples it may be seen that differentiation and integration are related in that flow is the derivative of volume and volume is the integral of flow.

$$\text{flow} = d(\text{volume})/dt$$

$$\text{volume} = \int_0^t \text{flow}\, dt$$

but if the volume that we are forming by our integration operation has a value at zero

$$\text{volume} = \left(\int_0^t \text{flow}\, dt\right) + \text{I.C.}$$

where I.C. is the initial condition, the volume at time zero.

Example 3

What is the integral of the function $f(t)$ shown in Figure 15(a)?

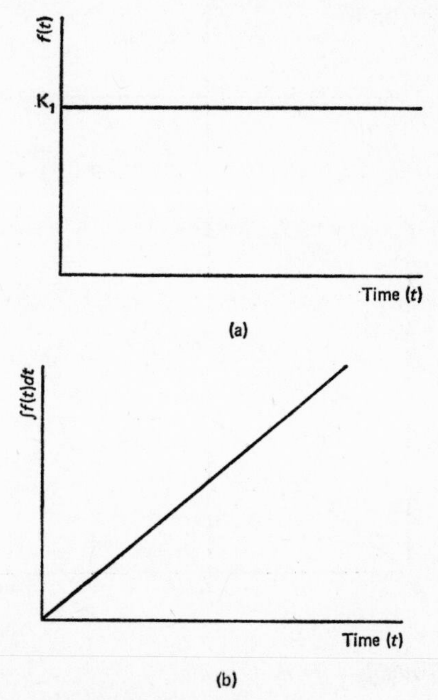

Figure 15 See Example 3 for description.

During each interval of time Δt the function $f(t)$ is constant. Therefore the integral increases at a constant rate, as shown in (b). If (a) represents flow into a container then (b) is volume. If (a) is speed then (b) is distance moved.

Thus $$\int_0^t K_1 \, dt = K_2 t$$

where K_1, K_2 are constants.

Example 4

What is the integral of the function shown in Figure 16(a)?

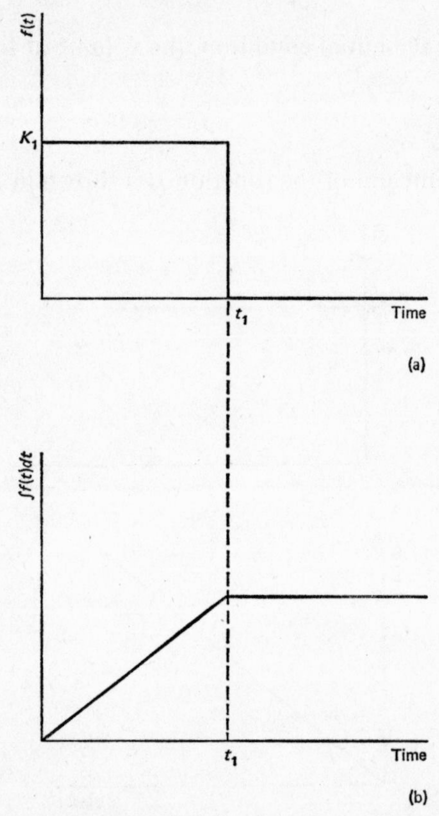

Figure 16 See Example 4 for description.

Until time t_1, $f(t)$ is identical to that shown in Figure 15(a), so the integral is also identical to Figure 15(b) up to this point. After time t_1, $f(t)$ becomes zero and so we continue to add zero to the integral, in other words, $f(t)$ becomes a constant. If the flow into a container suddenly stops, the volume in the container remains constant. If a car's speed drops to zero then its position remains constant.

DIFFERENTIAL EQUATIONS

There is a container of water with a tap at the bottom. The volume of water in the container is x, and therefore the flow through the tap is dx/dt, the rate of change of volume with respect to time. If we were to open the tap and note the emptying characteristics of the container we would find that flow out is proportional to the volume remaining. Thus

$$dx/dt = -Kx$$

where K is a constant. The negative sign indicates that rate of change is negative, i.e. a loss of volume. Figure 17 illustrates the

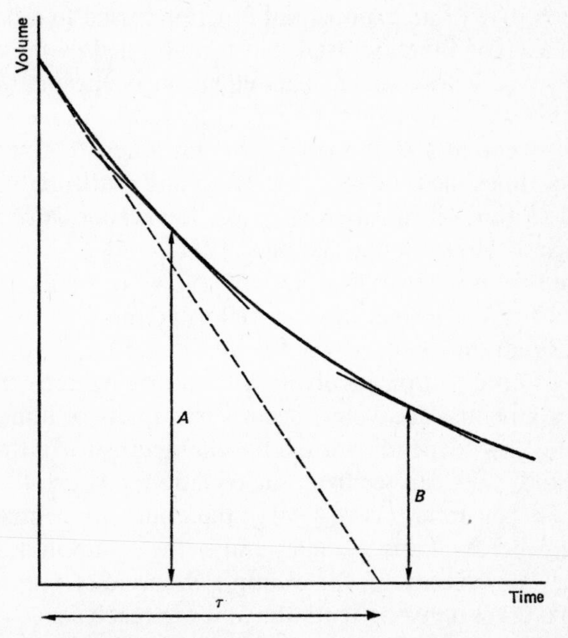

Figure 17 An exponential decay (see text for explanation).

volume in the container after the tap is opened at time zero. The rate of change of volume is proportional to the volume remaining, in other words the slope of the graph is proportional to its height. Distance B is one half of distance A and the slope at B is one half that at A. The rate of change becomes extremely small as the function approaches zero, and in theory the volume reaches zero only at time infinity. A slope of the kind shown in Figure 17 is known as an exponential decay, $\exp(-at)$. This refers to a function whose rate is proportional to its magnitude and the mathematics formalises this observation.

In solving the equation $dx/dt = -Kx$ it is clear that x must be a function whose derivative is merely a constant multiplied by the function. Such a function is $x = \exp(at)$ where exp refers to the base of natural logarithms (e) or the exponential, a is a constant and t is time, since the derivative of this function $dx/dt = a \exp(at)$. In the present case if $x = (1/K) \times \exp(-Kt)$, then $dx/dt = (-K/K) \times \exp(-Kt) = -\exp(-Kt)$ which satisfies the equation and so $x = (1/K) \times \exp(-Kt)$ is the solution. The derivative of an exponential function raised to a power t is the same as the function itself, but is multiplied by the constant that appears before the t. This equation is the mathematical description of the function shown in Figure 17. A function of the form $x = \exp(at)$ is the only function whose derivative is a constant times the function. The proof and mathematical background to this will not be presented, the reader can refer to a text such as Herstein and Sandler (1971).

Since flow is a constant times volume, with a change of scale Figure 17 can represent flow as well as volume.

The equation $dx/dt = -Kx$ is an example of a differential equation, an equation involving a term and its derivative.

An exponential decay may occur very rapidly or it may take a very long time, depending upon the characteristics of the system under study. We can see from the equation $x = \exp(-at)$ that if the constant term a is very large then our exponential decays rapidly, whereas if a is a small number the equation describes a relatively slow decay. If, for example, we were to multiply a by 10 then it takes only one-tenth the time t to reach a given point of decay as before.

In order to describe quantitatively the rate of decay of an exponential, we need some accepted standard. One such is the half life, i.e. the time a decaying function takes to reach the half way point between where it starts from and zero. However, in control system parlance we use a different measure, the 'time constant'. This may be explained by reference to Figure 17. The dotted line shows the rate of change of the function at time zero, and where this line crosses the zero line is the point in time at which the function would be zero were it to continue decaying at the original rate. The time τ, known as the time constant, is therefore the time the function would take to reach zero if the original rate were maintained. If we project this point upwards until it crosses the actual function $\exp(-at)$ we obtain a measure of the function at the time τ. In fact if we were to take such a measurement at time τ we could show that the exponential will have fallen by approximately 63 per cent of the distance to zero. Therefore the time constant may alternatively be defined as the time necessary for the exponential to complete 63 per cent of its journey.

In theory, at least exponentials decay to zero only at time infinity. However for all practical purposes we generally say that it has decayed sufficiently far after four time constants that we can ignore its presence.

Equations of Physical Systems

Figure 9 showed a spring with mass attached. If we apply a force F to the spring and note the displacement x we would find a linear relationship between force and displacement.

Thus $$F = K_1 x$$

where x = displacement, F = force, and K_1 = a constant. This is the defining equation for the system shown in Figure 9, which may be applied provided that the spring is not stretched beyond a certain limit where its mechanical properties would be altered.

Another basic mechanical component is shown in Figure 18(a). A mass is being pushed across a surface against the force of friction. We will ignore the various non-linear friction components that may be present, and take into account only the friction term which offers a resistance to motion that is propor-

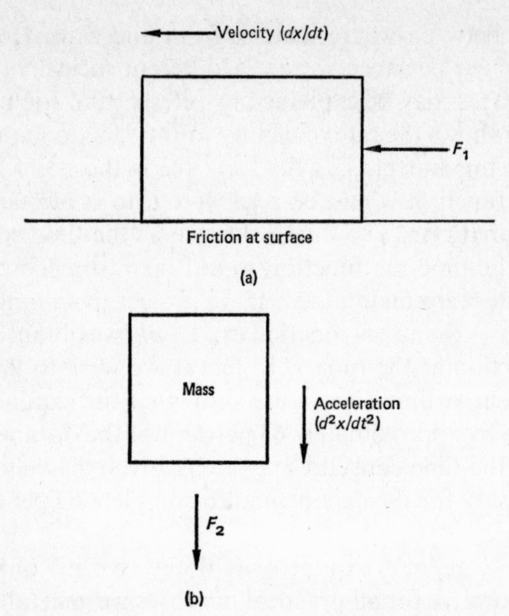

Figure 18 Two basic mechanical components, (a) friction and (b) mass.

tional to velocity. Since velocity is the derivative of displacement, the equation defining the frictional resistance is

$$F_1 = K_2 \times dx/dt$$

Figure 18(b) shows a mass which is being acted upon by a force, no friction being present. It could be a mass moving in space under a gravitational force, to give an example. According to Newton's law of motion the mass will experience an acceleration that is proportional to the force. Acceleration is the derivative of velocity with respect to time and velocity is the derivative of distance, so the defining equation for the mass is

$$F_2 = K_3 \times d^2x/dt^2$$

Having established the force components, we are in a position to consider the equation of a mass being accelerated where friction is present. Part of the applied force serves to overcome friction, the remaining part gives an acceleration to the mass. The total force is the sum of these components. Thus

$$F_1 + F_2 = K_2 \times dx/dt + K_3 \times d^2x/dt^2$$

Figure 19(a) shows a system consisting of a spring and dashpot. We assume that the mass of the structure is small enough to be ignored. The dashpot is a means of providing a frictional

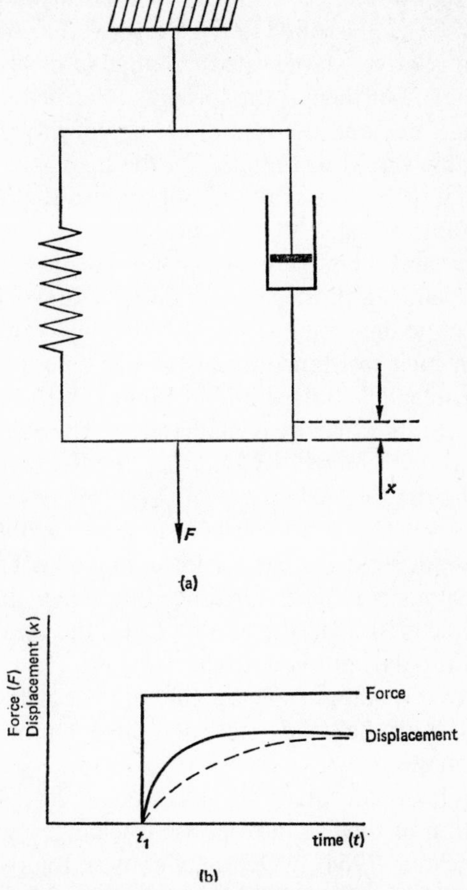

Figure 19 A system consisting of a spring and dashpot. (a) The mechanical structure, (b) the response of the system to a step change in force.

force that is proportional to velocity and its defining equation is therefore force = (velocity) × (a constant). We measure the displacement x from the unloaded position of the spring in response to the application of a force F. The equation of the system is

$$F = K_4x + K_5 \times dx/dt$$

K_4 is the constant of elasticity of the spring, while K_5 defines the force–velocity characteristics of the dashpot.

Let us suppose that we were suddenly to apply a force to the spring at time t_1. The force applied and the resultant displacement would be as illustrated in Figure 19(b). The force would at first cause a relatively large rate of change of displacement with respect to time, but then as the spring is stretched it would exert an opposing force and the rate of change of displacement with respect to time would be reduced. In the steady-state the elastic stretch of the spring is equal and opposite to the force applied. It is important to realise that the dashpot only offers resistance to movement and therefore can exert no influence upon the final steady-state value of displacement that is obtained. The dashpot can only change the speed with which the final value is reached, if it offers a high resistance to movement then a slow velocity response results. The dotted line in Figure 19(b) shows the response that results when K_5 is made large. The steady-state displacement is determined by K_4, i.e. the force–displacement characteristics of the spring.

Essentially we may consider the response to be the sum of two distinct components: a steady-state component for which the spring is responsible and a dynamic component that is defined by the characteristic of the dashpot. In the steady-state the presence of the dashpot is not felt.

The task that we face in solving differential equations is to find a solution to both the steady-state and the dynamic components of the response.

These fundamental equations of mechanics have been applied to the rotation of the eye in response to changes in target position (Westheimer, 1954), which may console the reader if he or she is wondering what possible relevance all of this has to biology.

It is perhaps not surprising that the principles of mechanics can be applied to the rotations of the eye, since we are dealing with a mechanical system involving elasticity of the muscles and possibly inertia of the eyeball. However it is not necessary for two systems to share a similar physical embodiment for their performance to be comparable. It is possible for a hormonal transport system in the body to be compared with an electric circuit, to give one example. In fact, biological subsystems are very commonly compared with electric circuits. The biologist may for instance speak of the resistance or capacitance of a system. This of course makes it essential to understand what is meant by such terms. In giving an explanation, an analogy will be drawn between electric current and the flow of water. One reason for doing this is that in all probability it is easier to understand invisible phenomena such as electric current and voltage by comparison with tangible terms such as water flow. But perhaps even more important, it introduces the idea of an analogy between systems having a different physical realisation. From our point of view the equations that characterise a system are more important than the physical embodiment of the system.

Figure 20(a) shows a U-tube with an unequal height of water between the two sides. An aperture P divides the two compartments, but for the moment it is closed. As soon as the aperture is opened water flows from side A to side B since the difference in height creates a pressure difference between the two sides. The flow rate of water depends upon the size of the aperture, as the aperture is made larger it then offers less resistance to flow, and so the flow increases.

Figure 20(b) shows an analogous electric circuit. The switch is open, but as soon as it is closed an electric current flows. Electric current represents the movement of electric charge and may be compared to the flow of water in cm^3/min. In order for a current to flow there must be a driving force, in other words, a source of potential difference or a voltage. Such a driving force is analogous to the pressure, arising from the difference of water heights in the two arms of the U-tube, that forces water through the aperture. When we speak of the resistance of an electric circuit we mean the resistance that is offered to electric current flow, and this may be compared with the resistance that the aperture offers to the flow of water.

The equation of an electric circuit comprising a resistor and a voltage source is

$$\text{voltage} = \text{current} \times \text{resistance}$$

If the resistance is doubled then the current that flows in response to a given voltage is halved. Resistance is a purely dissipative

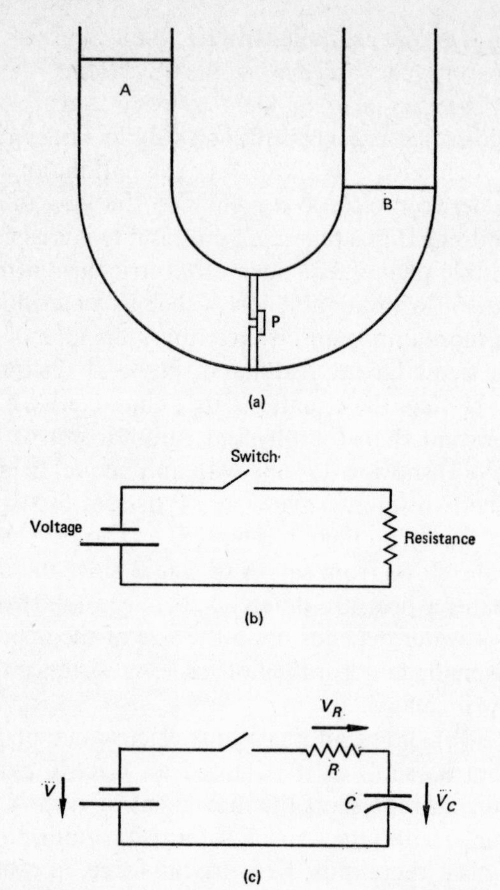

Figure 20 Analogous systems. (a) The difference in height of water on the two sides creates a force that attempts to move water through the aperture P, *(b) an electric circuit in which the voltage causes a current to flow through the resistance when the switch is closed, (c) a circuit consisting of a resistance and a capacitance.*

element, that is to say energy is expended in passing an electric current through a resistor. There can be no storage of energy in a resistor, and in this sense it may be compared to friction in a mechanical system.

Electric current consists of the movement of charged particles, charge being the integral of current with respect to time. If we apply an electric current to a component known as a capacitor, then charge will be stored. A capacitor consists of two plates separated by an insulating material, and in charging a capacitor electrons are applied to one plate and removed from the other. The consequence of this is that a voltage appears across the capacitor. If we charge a capacitor and then leave it unconnected (meaning an infinite resistance between its plates) the charge will remain on its plates, and consequently it will retain its voltage. However, if a resistor is connected across the capacitor, electrons are able to flow from one plate of the capacitor to the other through the resistor. The current will continue to flow until the charge on the capacitor and consequently its voltage is reduced to zero.

The voltage developed across a capacitor is proportional to the charge on its plates. Thus

$$Q = \int I \times dt$$

where Q is charge, I is current and t is time, and

$$V = KQ$$

where V is voltage and K is a constant. Actually K has the value $1/C$ where C is the capacitance of the capacitor, a fixed property of its construction. Thus

$$V = \frac{Q}{C}$$

Figure 20(c) shows a circuit consisting of a voltage, a resistor and a capacitor. The sum of the voltage that appears across the resistor V_R and that across the capacitor V_C is equal to the applied voltage V. This circuit enables the essential difference between a resistor and a capacitor to be understood. Let us assume that there is a constant voltage source that is applied to the circuit when the switch is closed. Current will flow until the charge on the capacitor is equal and opposite to the applied voltage, at which point the current is zero. Suppose that before the capacitor were fully charged we were to open the switch so as to prevent further current from flowing in the circuit. The voltage across the resistor V_R would immediately drop to zero, since this voltage is given by the product of current and resistance. However, since the voltage across the capacitor is proportional to the integral of the current that has been flowing up to the time the switch was opened, this voltage remains at the value it had immediately prior to opening the switch.

When a voltage is applied to the circuit shown in Figure 20(c), the current flows until the voltage across the capacitor is equal and opposite to the applied voltage. When the current is zero the steady-state has been reached, and the resistor has no influence on the steady-state solution. However, the resistor has a profound influence on the speed with which the steady-state is reached. An electrical resistor is therefore analogous to the dashpot in the mechanical system shown in Figure 19(a). In the mechanical system the steady-state is reached only when the force exerted by the spring is equal and opposite to the force that is applied from outside the system. The speed with which the steady-state is reached is determined by the dashpot. We are therefore dealing with two analogous systems, the one mechanical and the other electrical.

Linearity and Non-Linearity

To take a practical example, a force of 1 kg is applied to a mass and the acceleration that results is 1 kg/s². The force is doubled and the acceleration doubles. The force is increased tenfold and the acceleration is found to be ten times as great. From these observations we would be able to say that the system is linear, at least over the range tested. The two variables, force and acceleration, are related throughout by the same constant parameter K_3, i.e. $F = K_3 \times d^2x/dt^2$.

If we were to apply increasing force to the spring shown in Figure 9, we might find, for example, that 1 kg would produce a displacement of 1 cm, and 5 kg a displacement of 5 cm. Within a certain limit the system is linear. However, before long we would find that displacement increased at a lower rate with increases in force, and 7 kg may only give a displacement 6·2 cm instead of 7 cm. We would then have exceeded the linear operating range of the spring. Figure 21(a) shows the behaviour of the spring. The dotted line represents the linear relationship $F = K_1x$, while the solid line shows the actual behaviour of the spring which may be seen to violate the equation when a large

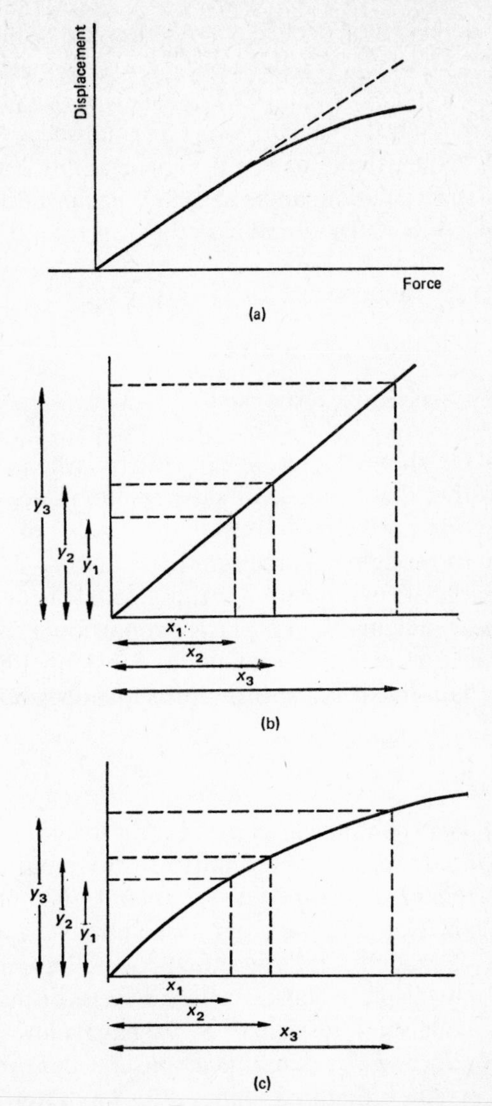

Figure 21 Linearity and non-linearity. (a) A non-linear relationship between displacement and force with the linear function shown dotted, (b) a linear function, (c) a non-linear function. (For explanation see text.)

force is applied. The system is linear for inputs within a range, but at the upper limits it is non-linear and an equation having a fixed parameter cannot describe it over the whole range.

A linear dependence between two variables is illustrated in Figure 21(b), where the response y_1 is produced by the input x_1 and the response y_2 by the input x_2. What happens if we apply an input x_3 that has the value $x_1 + x_2$? The output y_3 that results may be seen to have the value $y_1 + y_2$. This example illustrates the principle of superposition, and a system is linear if this principle may be applied to it. To summarise the principle:

If y_1 is the response to x_1

y_2 is the response to x_2

then $y_1 + y_2$ is the response to $x_1 + x_2$

Figure 21(c) shows a non-linear system, where it may be clearly seen that the principle of superposition does not apply. In this case, the output y_3 in response to x_3 is not the sum of y_1 and y_2 even though x_3 is the sum of x_1 and x_2.

The principle of superposition forms a restatement of the fact that for linear systems the output is proportional to the input (for example, acceleration equals a constant times force) whereas for non-linear systems the principle does not apply.

Gain

The word *gain* is employed to express the relationship of the output amplitude of a system relative to the input amplitude. Let us refer again to Figure 1 in Chapter 1. The output, displacement y, is related to the input, displacement x, by the gain R_2/R_1. If R_2 is made much larger than R_1, we have a high gain in that a small displacement x causes a large displacement y, but if R_2 is made small relative to R_1 we have a low gain. If we fix values for R_2 and R_1, then it is a constant term R_2/R_1 – the gain of the system – that determines y for any given x. Gain is commonly written as K.

For linear systems, such as that shown in Figure 1, gain is a constant but for non-linear systems gain may vary as a function of the amplitude of the input. The spring shown in Figure 9 has a gain of 1 cm per kilogram for small loads but its gain is less as the load is increased.

A GAIN OR A GAIN PLUS LAG

Let us return to the lever shown in Figure 1. If we cause a displacement x at the input end we obtain a displacement y at the output end, the relation between y and x being $y = (K_1/K_2)x$. We have just noted that y is proportional to x for all values of x, in other words we have a linear system. What is also true is that the transmission of 'information' in the system is instantaneous. What is meant by this is that if the lever is made of perfectly rigid material, y does exactly what x does without delay or distortion. A graph of y as a function of time would look just like a graph of x as a function of time, except for the amplitude change determined by the ratio K_1/K_2.

Contrast this with the spring and dashpot shown in Figure 19. Here we apply a force, and the pointer on the spring moves to a new position. However, unlike the lever, it does not execute its response immediately, but after the sudden application of a force it approaches its new position exponentially. Although we might, after measuring the final displacement, note that the system has a gain of, say, 1 cm per kilogram, this information is insufficient to characterise the system completely. In addition, we need to specify how quickly it responds following the application of a force. To do this we use the concept of the time constant. If the pointer takes 1 min to complete 63 per cent of its journey we would say that it has a time constant of 1 min. A gain in the steady-state of 1 cm per kilogram and a time constant of 1 min has now completely characterised the system. A component whose response is an exponential rise to a steady value, following a sudden increase in the stimulus to a steady value, is known as an exponential delay or lag. Lags of this kind are found as components in very many systems.

Turning our attention again to the lever, let us suppose that it is hidden from us, and we are allowed to see only the input displacement x and the output displacement y. We now apply various test stimuli (called the input) and observe the response (called the output). This is the so called black box approach, and from the input–output information we attempt to characterise the black box. We could, for example, cause a sudden instantaneous displacement x which we would call a step input.

For this system the output would also be a step of magnitude K_1/K_2 times the input. If now a sinusoidal input is applied to the lever by causing displacement x to vary up and down such as to describe a sinewave, we would find that the output would also be a sinewave, having an amplitude K_1/K_2 times the input. This would be true whatever frequency of sinewave we applied, and we would be justified in characterising the black box as a pure gain.

We could now do the same thing to the spring and dashpot shown in Figure 19. We know that for a step input of force we obtain an exponentially delayed output. Suppose we now apply a very slow sinewave input to the system. The displacement of the pointer will also vary sinusoidally through a particular amplitude. If we now make the input sinewave have a higher frequency of oscillation, the output will still oscillate sinusoidally but the amplitude of the oscillations will be less than when a low frequency was applied. This is the behaviour of a system with a lag in it and, therefore, reinforces the conclusions drawn from the response to a step input.

Why does the amplitude of the sinewave get smaller as we increase the frequency? We have seen from the response to a step input that the system cannot respond immediately. If we apply a very slow sinewave there is sufficient time for the system to follow it. However, if we increase the frequency of the input sinewave, no sooner has the output started to move in one direction than it is called upon to move in the opposite direction. The result is a small oscillation at the input frequency. If we increase the frequency still more then the output has no chance to move before a reversal of the stimulus occurs. It therefore shows no oscillation. We will look at the detailed frequency response of systems later, and will predict the output to be expected for particular input frequencies.

Is the spring and dashpot an example of a linear system? Yes, provided that signals of reasonable amplitude are applied. Thus if we apply a step x_1 and obtain an exponential response y_1, and apply a step x_2 and obtain an exponential response y_2, a response y_1 plus y_2 will be obtained for a stimulus x_1 plus x_2. That is the only requirement for a system to be linear.

3 THE LAPLACE TRANSFORM METHOD

The Laplace Transform

In the last chapter we discussed differential equations, and as one example considered the equation that defines a particular mechanical system. The solution of such equations involves finding what happens to, for example, displacement after the application of a particular force. We require displacement x as a function of time t, and this involves solution of an equation that includes terms such as dx/dt. There are traditional methods available to us for solving differential equations, but we will completely by-pass them in this book. Instead we will use the much simpler technique of the Laplace transform.

This technique involves the transformation of a differential equation into an algebraic equation. When using the Laplace transform, both the transient and steady-state components of the response appear in the solution.

Only the practical usage of the method is described here; the reader can find the mathematical background to the method in one of the numerous texts on this subject. We merely use the technique as a tool in order to solve problems.

The Laplace transform of a function $f(t)$ is written as $\mathscr{L}[f(t)]$ and is defined by the equation

$$\mathscr{L}[f(t)] = \int_0^\infty f(t) \times \exp(-st)\, dt$$

s being a complex number.

Though the analogy perhaps should not be taken too far, the process of taking the Laplace transform of $f(t)$ can be compared to that of taking the logarithm of a number. Thus when we write $\log_{10} 2 = 0{\cdot}3$, the number 2 has been transformed into its logarithm to the base 10 and the result 0·3 has been obtained. We have transformed from one domain into a second domain.

$\mathscr{L}[f(t)]$ represents the process of transformation from the time to the Laplace domain. Each number has a particular and unique logarithm, and similarly the various time functions that we will deal with each have their unique Laplace transform. In neither case do we lose any information in making the transformation

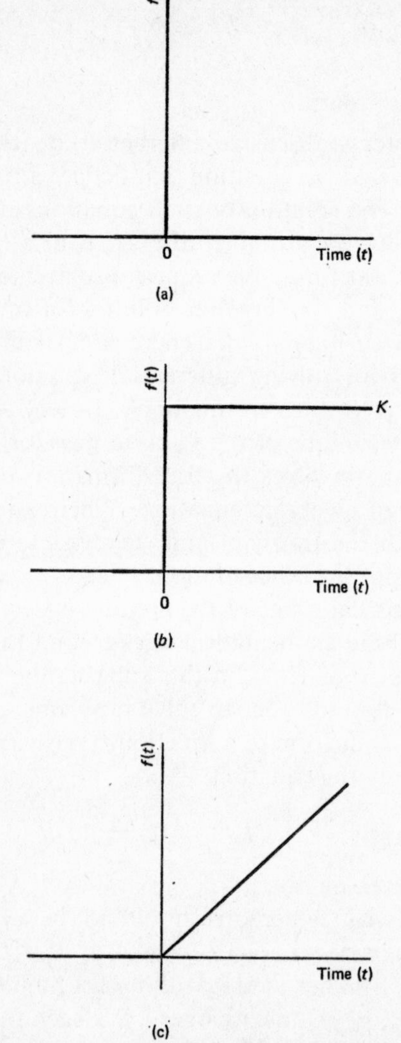

Figure 22 The three most common test signals. (a) The impulse, (b) the step, (c) the ramp.

from one domain to another, because the unique identity is preserved and we can always return to the original domain. In this sense it is not like translating from English to German. We convert numbers to logarithms and perform our calculations upon logarithms simply because the mathematics is made easier. Conversion of a differential equation into an algebraic equation involving Laplace is done for the same reason. The conversion of $f(t)$ into $\mathscr{L}[f(t)]$ is represented by $F(s)$. $F(s)$, sometimes written as $\bar{f}(s)$, is the Laplace transform of $f(t)$.

At this point it is necessary to consider some of the time functions that we normally encounter in using control systems techniques, and then to consider how to transform these time functions into their Laplace equivalents. Three time functions are shown in Figure 22. Part (a) illustrates a unit impulse, that is defined as $f(t) = \infty$ for $t = 0$, and $f(t) = 0$ for $t \neq 0$. It is perhaps somewhat difficult to imagine a function that suddenly rises to infinity at time zero and then immediately returns to zero again. To give a very crude approximation to such a function we can take the case of the stimulation of the visual system by light. If we were to sit a subject in a completely dark room, and at time zero fire a photographer's flash bulb, we would obtain a signal of high amplitude and short duration.

Figure 22(b) shows a step function that is defined as $f(t) = K$ (a constant) for $t > 0$ and $f(t) = 0$ for $t < 0$. To return to our practical example, this would describe the stimulus produced by a light of constant intensity being switched on at time zero and kept on. Figure 22(c) shows a ramp function, $f(t) = Kt$, in which the light intensity starts at zero and increases linearly as a function of time.

It is easiest to take the step function in order to illustrate how to find the Laplace transform of a function. In this case $f(t)$ takes the constant value K.

$$\begin{aligned}\mathscr{L}[f(t)] &= \int_0^\infty f(t) \times \exp(-st)\, dt \\ &= \int_0^\infty K \exp(-st)\, dt \\ &= \left[-\frac{K \exp(-st)}{s}\right]_0^\infty\end{aligned}$$

when $t = \infty$

$$-\frac{K \exp(-st)}{s} = 0$$

when $t = 0$

$$-\frac{K \exp(-st)}{s} = -K/s$$

$$\mathscr{L}[f(t)] = 0-(-K/s)$$

$$F(s) = K/s$$

The reader will probably be relieved to know that it is not necessary to go through such a calculation for every time function that is likely to be met. Tables of these are available, and transformations of some of the most common time functions appear in Table 1.

TABLE 1. *Laplace transformations of some common functions*

Time function	*Laplace transform*
The unit impulse $\partial(t)$:	
$f(t) = \infty$ for $t = 0$	1
$f(t) = 0$ for $t \neq 0$	
The step function:	
$f(t) = K$ for $t > 0$	K/s
$f(t) = 0$ for $t < 0$	
The ramp function:	
$f(t) = Kt$ for $t > 0$	K/s^2
$f(t) = 0$ for $t < 0$	
The decaying exponential:	
$f(t) = \exp(-at)$	$1/(s+a)$

One of the basic rules of Laplace transform manipulation is the principle of linearity. If $\mathscr{L}[f(t)] = F(s)$ then $\mathscr{L}K \times f(t) = K \times F(s)$. It follows also that

$$\mathscr{L}[f_1(t)+f_2(t)] = \mathscr{L}[f_1(t)]+\mathscr{L}[f_2(t)]$$

It may be proven that if the Laplace transform of $f(t)$ is $F(s)$ then the Laplace transform of $df(t)/dt$ is given by $s \times F(s)-f(0^+)$

where $f(0^+)$ means the value of $f(t)$ at time zero as t approaches zero from the positive side. For the purposes of control systems analysis it is assumed that the initial conditions of the time functions that we deal with are zero. If a function is zero before time zero, and then at time zero takes some value other than zero – which is the case for the step function – then the function is not defined at time zero. In practice we get what we believe is the correct solution by ignoring the initial conditions, and though this is not very satisfying from a mathematical point of view it works, and so it is employed. If one uses the rigorous mathematical definition, then one obtains a solution for all times other than time zero, at which point the function is not defined. In other words if we assume that the initial conditions of a function are zero, then in order to get the Laplace transform of the derivative of a function we multiply the Laplace transform of the function by s. It really is as simple as that, and it works.

In the case of integration

$$\mathscr{L}\left[\int_0^t f(t)\,dt\right] = \frac{1}{s} \times F(s) + \frac{\left[\int f(t)\,dt\right]_{t=0}}{s}$$

where $F(s)$ is the Laplace transform of $f(t)$. It is somewhat clearer what is meant by initial conditions in this case. We mean the value of the integral at time zero when we start our calculation; practical examples are the volume of fluid in a container or the charge on a capacitor. In this case there is no ambiguity, if the integral has a value at time zero we must take it into account and there is no problem in defining what is meant by the value at zero. However, in control system applications the initial conditions usually are zero. If this is the case then it is remarkably easy to obtain the Laplace transform of the integral of a function $f(t)$ since we merely multiply the Laplace transform of $f(t)$ by $1/s$.

The Inverse Laplace Transform

If a function of time $f(t)$ has a Laplace transform $F(s)$, then conversely the Laplace transform $F(s)$ has a 'time transform' $f(t)$. In fact, $f(t)$ is known as the inverse Laplace transform of $F(s)$, and is written as $\mathscr{L}^{-1}[F(s)] = f(t)$. This may be compared to the anti-logarithm, and Figure 23 attempts to make the operation clearer. In the case of logarithms, we multiply num-

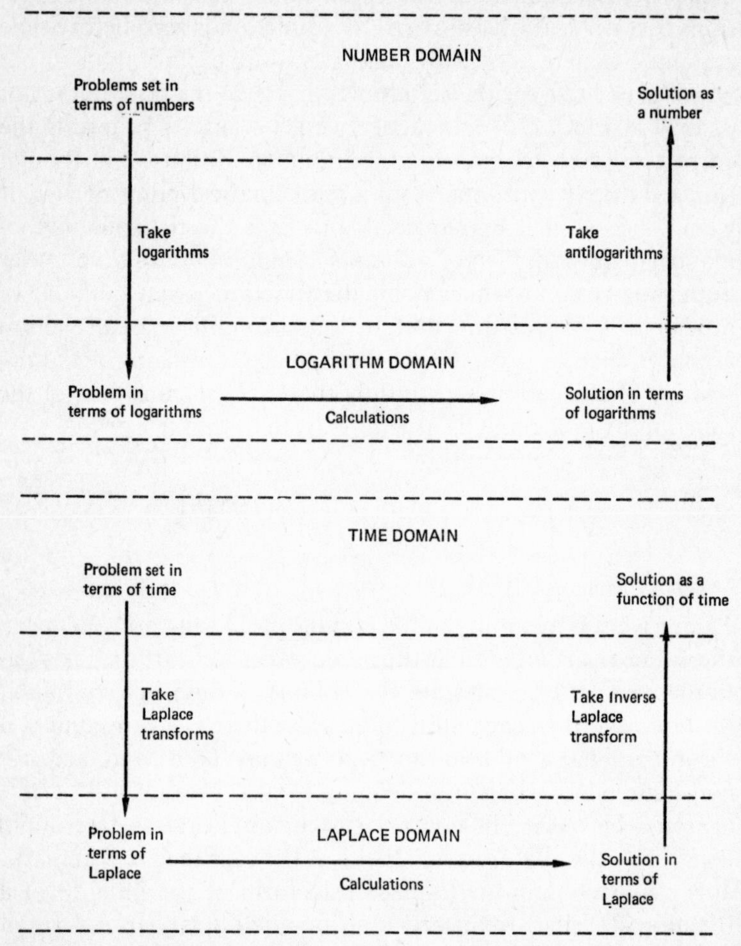

Figure 23 A comparison between the process of solving a problem by using logarithms and solving a time dependent equation by means of the Laplace transform. In each case we transform to a new domain, do our calculations, and then convert back into the domain of interest.

bers by adding logarithms and obtaining the logarithm of the product. In order to obtain the result of the multiplication operation, we look up the result of our addition in a table of anti-logarithms. In other words, we transfer from one domain in order to do our calculations and then, having obtained the result, we relate this result to the domain of interest.

In the case of Laplace transforms, after we have obtained the transform of the solution to our problem, we then require the inverse transform, i.e. the time function corresponding to this Laplace transform. If we are lucky, all that is involved in finding the inverse transform is to look up the Laplace transform in a set of tables, a part of which is shown in Table 1. We observe which time function corresponds to the Laplace transform that we have obtained. Sometimes the Laplace transform will not appear in the tables, in which case it must be broken up into a sum of simpler terms each of which appears in the tables. For example, let us suppose that we want to look up the inverse Laplace transform of $1/[s(s+1)]$. The technique of partial fractions may be employed to reduce the expression to a sum of simpler terms. Thus

$$1/[s(s+1)] = A/s+B/(s+1)$$

A product as a denominator is equal to the sum of the terms forming the denominator, A and B being constants. We can multiply both sides of the equation by $s(s+1)$ in order to simplify it. Thus

$$1 = A(s+1)+Bs$$

$$1 = As+A+Bs$$

We now work on the basis that the coefficients of s^n on the left side of the equation are the same as those on the right side. Let us consider the case when $n = 0$, i.e. we consider the constants on each side. The constant on the left-hand side is 1 and on the right-hand side is A. If constants are equated on the two sides, the result is obtained that A is 1. Considering s^1, there are no terms involving s^1 on the left-hand side, but on the right-hand side there are $A+B$ terms as the coefficient of s^1. Therefore

$$0 = A+B$$

and since $A = 1$

$$B = -1$$

therefore, the expression

$$1/[s(s+1)]$$

may be replaced by

$$1/s-1/(s+1)$$

Each of the components $1/s$ and $1/(s+1)$ appears in Table 1. Therefore, inverting each of them and forming their sum

$$\mathscr{L}^{-1}1/s(s+1) = 1-\exp(-t)$$

If we were given the function $1/s^2(s+1)$ to invert, again we would break it up into a sum of simpler terms in the following way

$$1/s^2(s+1) = A/s^2+B/s+C/(s+1)$$

The term B/s arises because, where a term raised to a power n is present (in this case A/s^2), then $s^n, s^{n-1}, s^{n-2}, \ldots, s^1$ must also be included.

$$1 = A(s+1)+Bs(s+1)+Cs^2$$
$$1 = As+A+Bs^2+Bs+Cs^2$$

Equating s^0 on each side

$$1 = A$$

Equating s^1 on each side

$$0 = A+B$$
$$B = -1$$

Equating s^2 on each side

$$0 = B+C$$
$$C = 1$$
$$1/s^2(s+1) = 1/s^2-1/s+1/(s+1)$$

Each of these terms appears in the Laplace transform tables

$$\mathscr{L}^{-1}1/s^2(s+1) = t-1+\exp(-t)$$

The use of the Laplace transform is best illustrated by means of working through examples, some of which now follow.

Example 5

A mass m has a force applied to it; the force rises from zero to a constant value K at time zero and then remains there. The velocity of the mass is zero at time zero. What is the velocity of the mass after application of the force?

The equation that defines the system is Newton's law of motion that force f equals mass m times acceleration a.

$$f(t) = m \times a(t)$$

Acceleration is the derivative of velocity v with respect to time

$$f(t) = m \times dv/dt$$

The Laplace transform of $v(t)$ we can call $V(s)$ and so, according to the rule of differentiation, the Laplace transform of dv/dt is $s[V(s)]$. As m is a constant, we are able now to transform the equation by taking the Laplace transform of each side. Thus if

$$f_1(t) = f_2(t)$$

then $$F_1(s) = F_2(s)$$

so it follows that since

$$f(t) = m \times dv/dt$$
$$F(s) = m \times s[V(s)]$$

$F(s)$ is given by K/s since the Laplace transform of a step K is K/s.

$$K/s = m \times s[V(s)]$$
$$V(s) = \frac{K}{ms^2}$$

If we inspect Table 1 we will find that the Laplace transform of a ramp function Kt is K/s^2. In the present case we have a function of the form of a constant divided by s^2, the constant being K/m. Therefore, taking the inverse Laplace transform we get

$$v(t) = \frac{K}{m} \times t$$

It is in some way a little unfortunate that we have two Ks that mean different things, but since the reader is likely to meet this in practice, I have deliberately not simplified it by calling one of them K_1. The symbols are of course quite arbitrary, it is only important to understand that in the Laplace tables K means the constant that multiplies the time or Laplace function. In our example the constant that multiplies the term $1/s^2$ is K/m.

Example 6

Figure 24 shows a circuit consisting of a voltage source E, a resistance R and a capacitance C, all of which have constant values. The charge on the capacitor is zero at the time the switch is closed, that we will call time zero. What is the current flow (i) that occurs after the switch is closed?

The first stage is to consider the voltage that will appear across each of the two components when a current flows. The equation of a resistor is that voltage equals resistance times current. In other words $v(t) = i(t) \times R$. The equation for a capacitor is

$$\text{voltage} = \int \frac{i \times dt}{C}$$

The voltage is proportional to the integral of current with respect to time and inversely proportional to the capacitance C. The sum of the voltage across the resistance and that across the capacitance will be equal to the voltage that is applied. This enables the equation of the system in the time domain to be written.

$$E(t) = i(t) \times R + \int \frac{i(t) \times dt}{C}$$

Taking Laplace transforms of both sides of the equation

$$\mathscr{L}E(t) = \mathscr{L}[i(t) \times R] + \mathscr{L}\left[\int \frac{i(t) \times dt}{C}\right]$$

$E(t)$ is a step input that therefore has the Laplace transform E/s. $I(s)$ represents the Laplace transform of $i(t)$ and this forms the unknown that we are trying to find.

$$E/s = I(s) \times R + I(s)/(sC) + \frac{\left[\int i(t) \times dt\right]_{t=0}}{sC}$$

The charge on the capacitor, in other words the initial conditions of the integration operation, are zero. This means that the expression reduces to

$$E/s = I(s) \times [R+1/(sC)]$$

$$I(s) = \frac{E}{s[R+1/(sC)]}$$

Looking at Table 1 it may be seen that if this expression can be converted into the form $K/(s+a)$ it may then be inverted

$$I(s) = \frac{E/R}{s+1/(RC)}$$

$$i(t) = \frac{E}{R}(\exp[-t/(RC)])$$

When t is zero $\exp[-t/(RC)]$ is 1 and therefore the current is given by E/R and is at a maximum. As from time zero the capacitor acquires a charge, since a current begins to flow, and this charge on the capacitor creates a voltage that serves to oppose the flow of current. Thus current flow $i(t)$ falls exponentially. As t approaches infinity, $i(t)$ will approach zero, in other words the current flows until the charge on the capacitor produces a voltage that is equal and opposite to the applied voltage. When this happens there is no driving force for the current and so it reaches zero.

Example 7

In Figure 24(b) we have the same circuit but this time there is a charge Q_0 on the capacitor at time zero, as indicated in the diagram. What is the current that flows in response to the same step input of voltage?

$$E(t) = i(t) \times R + \int \frac{i(t)\,dt}{C} + \text{initial voltage on the capacitor}$$

Taking Laplace transforms of both sides of the equation, and remembering to take the initial conditions into account

$$E/s = I(s) \times R + I(s)/(sC) + \text{initial voltage}/s$$

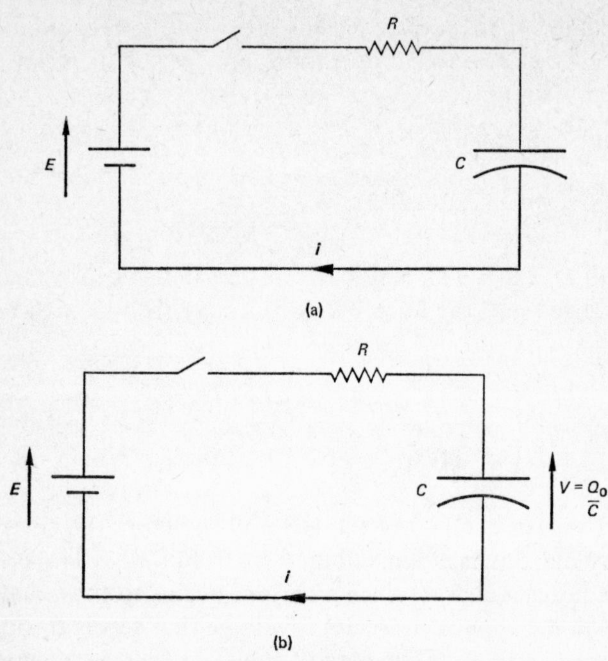

Figure 24 A resistance–capacitance circuit. (a) No charge is on the capacitor, (b) a charge is on the capacitor.

The initial voltage on the capacitor is given by Q_0/C

$$E/s = I(s) \times R + I(s)/(sC) + Q_0/(sC)$$

$$E/s - Q_0/sC = I(s) \times [R + 1/(sC)]$$

Multiplying each term by s/R

$$E/R - Q_0/(RC) = I(s) \times [s + 1/(RC)]$$

$$I(s) = \frac{E/R - Q_0/(RC)}{s + 1/(RC)}$$

This is now in the form $K/(s+a)$ and can readily be inverted

$$i(t) = \frac{E - Q_0/C}{R} \times \exp\,[-t/(RC)]$$

In this case the current at time zero is not given by E/R, since the driving force at time zero is not simply the applied voltage E, but rather is the difference between the applied voltage and the opposing voltage that arises from the charge on the capacitor.

4 MATHEMATICAL MODELS

A model is a representation of a system in a form that, while being different from the real system, none the less exhibits its essential features. The construction of a working model of a system is a quite familiar process. For example, a scale model of an airport may be only a few centimetres wide but still be able to illustrate the essential principles of the real system. With reference to such a model one could discuss aircraft movements and wind directions in a reasonably realistic way. The flow of information in such a model is in two directions. The model is based upon the real system, but at the same time the model generates new and unexpected information that refers to the real system.

We are concerned here only with a particular kind of model: the mathematical model. Such a model shows the essential structure of a system in the form of mathematical operators that characterise the performance of the system. Again, the flow of information is two way. The model is based upon a biological system, and tests are carried out on the model that serve to illuminate our understanding of the biological system. Therefore, the operation of model building is not to be seen as the passive copying of biological systems, but rather as a process of explanation usually carried out with a programme of experiments. Modelling suggests experiments that then serve either to confirm or challenge the validity of the model. In this way a parallel development between theory and experimentation is obtained.

As a starting point for the discussion of mathematical models, consider an object of mass M_1 which has a force F applied to it. The mass experiences an acceleration of value a_1. The applied force is doubled and the acceleration is found to double also.

The force is increased by a factor of 100 and an acceleration of 100 times a_1 is observed. As a result of this we might feel justified in writing the equation

$$\text{acceleration} = K \times \text{force}$$

where K is a constant. Having established a linear dependence between two observable terms, we might be content with the constant K, on the other hand we might prefer to attribute some physical significance to K. Consider now a second mass M_2, which is double M_1 and has the same force F applied to it. The acceleration a_2 which now results is only one half of a_1. In doubling the mass we have halved the acceleration, and so clearly the constant K that relates force and acceleration is dependent upon the mass. In fact the equation governing motion is

$$\text{force} = \text{mass} \times \text{acceleration}$$

or in our terms

$$\text{acceleration} = (1/\text{mass}) \times (\text{force})$$

The constant of the previous equation is now given some kind of physical realisation in that it is 1/mass, and is somewhat less of an arbitrary term designed to relate two observable events. At a simple level the equation forms a model of the real system, and one which corresponds directly to our understanding of inertia. Although we have not explained the inertial property of a mass we have at least anchored the equation to some measurable and identifiable property of the physical world. The example is perhaps somewhat unreal in the sense that the experimenter concerned knew nothing about Newton's work, and yet knew that mass had something to do with the amount of material in a body. However, the example is designed merely to illustrate the idea of model-building.

It will be emphasised later, when complex systems consisting of many components are analysed and, therefore, when many possible fits can be made, that models ought to be explanatory and not merely descriptive. That is to say their components refer to identifiable aspects of the real system. Does the model incorporate the essential operations that are necessary to produce the observed performance in the real system? A model of the force–acceleration dependence which includes mass as a component clearly goes as far as we possibly can towards satisfying this criterion. A constant K, even if we can give a number to K, does not necessarily refer to mass, except by implication. It is very easy to produce a mathematical model of a system in biology, for instance, but probably the most relevant question to ask is whether the model is based upon identifiable components of the real system.

Figure 25(a) represents a fundamental component of the circulatory system, a blood vessel. There is a pressure difference between the ends of the vessel and this difference in pressure

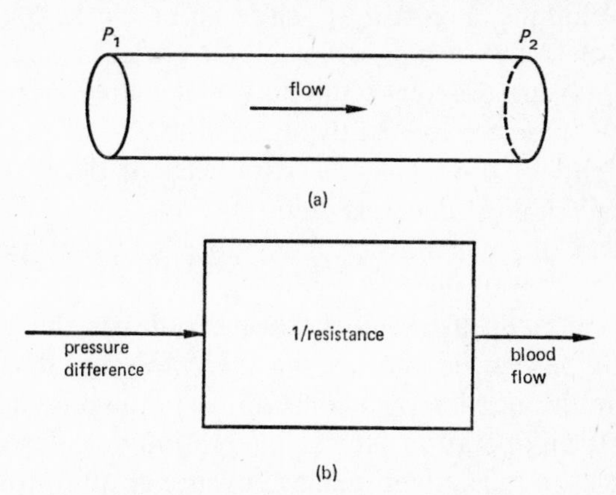

Figure 25 A blood vessel. (a) There is a pressure difference $P_1 - P_2$ between the ends which causes a flow, (b) a model of the pressure–flow dependence, showing that it is defined by resistance.

causes blood to flow from the high to the low pressure end. Let us assume that we can vary the pressure gradient ($P_1 - P_2$) and measure the blood flow which results. Within limits the flow would be found to vary linearly with the pressure difference.

$$\text{flow} = K \times \text{pressure difference}$$

K is dependent upon the characteristic of the vessel and is based upon its length and diameter. A variable called flow is dependent upon a variable called pressure, and the dependence is mediated via the blood vessel. K therefore takes on a particular significance as far as the physiology of the system is concerned. In practice, physiologists normally consider the resistance that a vessel offers to blood flow, there being a standard equation for resistance in terms of the dimensions of the vessel (see Guyton, 1969 for details). The relevant equation for flow and pressure is then

$$\text{flow} = (1/\text{resistance}) \times \text{pressure difference}$$

In this case K equals 1/resistance. Resistance is not an arbitrary parameter designed to match flow to pressure, but rather is a term that depends upon identifiable and measurable characteristics.

An alternative method for representing systems is the block diagram technique. Figure 25(b) shows a block diagram of the pressure–flow relationship of the blood vessel. There are two variables, pressure difference and flow, which are represented by arrows, and these are related by an operator 1/resistance which is the contents of the block, the parameter of the system. The variable called flow depends upon the other variable called pressure and this dependence is produced by the operator, a constant which multiplies an incoming signal to give an output signal. The content of the block always indicates the operation which is to be carried out on the incoming signal, and may consist of multiplication by a constant, as in the present case, or alternatively an operation such as integration or differentiation. The variables of the system such as flow are always shown as an arrow. One particular arrow corresponds to one particular variable. The mere presence of a constant term as in Figure 25(b), always indicates multiplication by that term. Differentiation would be given by d/dt and integration by $\int dt$. The contents of the block mathematically characterise the effect of the operator on the input signal. If we put a signal into a block and its derivative appears at the output then we mathematically characterise the operation as being one of differentiation and put d/dt in the block. This does *not* mean d/dt times the input.

The reader is probably wondering how realistic a model of the kind shown in Figure 25(b) is, and under what conditions it might be expected to represent what is happening in the physiological domain. If one refers back to the acceleration of a mass, it is apparent that this case was an example of a system describable under all conditions by the equation given (ignoring relativistic phenomena). Unfortunately, in biology life is seldom as simple as that.

Let us suppose we were able to increase the pressure gradient across the blood vessel and measure flow. The constant parameter 1/resistance would define the relationship over a range, but then resistance itself might change and we could not assume a fixed parameter. Finally the limit of the physiological system might be reached, and therefore flow would be unable to go on increasing. To take an extreme example, the wall of the vessel might rupture and we would be dealing with a pathological rather than physiological system. The presence of a test signal in a biological system is liable to change the nature of the system under study, and the analyst must always be on guard against this. After all, if a force were to break a mass into several pieces, then our measurements might be made more difficult.

To take another example, consider the thirst mechanism of an animal. If a moderate stimulus in the form of an injection of x mEq of sodium chloride were to result in y cm^3 of water being drunk we would find that $2x$ mEq would produce approximately $2y$ cm^3 of drinking. However, we might well find that $10x$ would produce absolutely no drinking, simply because the animal was made uncomfortable by the large excess of sodium in its body fluids.

These examples were made deliberately extreme in order to illustrate the point, but it must be emphasised that when constructing models there are all possible responses between exactly what is predicted, and complete lack of agreement between the simulation and the real system.

For another exercise in model-building we turn to the visual system and attempt, by means of the inclusion of a non-linearity, to make the model applicable under all conditions. The model is a simplification but none the less serves to exhibit the essentials of the real system.

The retina contains specialised components serving distinct functions and carrying out information extraction at this, the first stage of the visual pathway. Some retinal components cause firing in particular optic nerve fibres when the intensity of the light is increasing, but they are insensitive to constant illumination. Others cause firing only when the light intensity decreases, while a third type are activated by steady levels of illumination. We will consider a mathematical model of a component that fires only when the light intensity increases. If we can assume, as a first approximation, that within limits the activity is linearly proportional to the rate of increase of illumination, then the defining equations are

$$f = K \times \frac{dI}{dt} \quad \text{for } \frac{dI}{dt} \geqslant 0$$

$$f = 0 \qquad \text{for } \frac{dI}{dt} < 0$$

where f = rate of firing, I = illumination, t = time and K = a constant. The reason that two equations are needed is that the response is not symmetrical, for decreases in illumination there is no response.

Figure 26(a) shows a block diagram of the system. The input signal is I, which is the magnitude of illumination. This signal enters a block which carries out the operation of differentiation (d/dt) upon it to give the rate of change of illumination with respect to time (dI/dt), this then forms the input to a non-linear operator. The concept of non-linearity was introduced in Chapter 2, where it was explained that it implies that the operation carried out on a variable is dependent upon the magnitude of that variable. Thus the operation must be specified as a function of the variable which is being operated upon. The relationship between rate of change of light intensity, and firing rate in the optic nerve assumed for this component, is non-linear and is shown in Figure 26(b).

For each value of dI/dt there is a corresponding value of f. For all negative values of dI/dt, f takes the value of zero since the system is insensitive to decreases in illumination. For positive values of dI/dt below the value marked 1, then $f = K \times dI/dt$, K being the slope of the line. However, there is a limitation on the maximum rate of firing, f_{max}, which occurs at point 1. Increases in dI/dt above this value fail to produce any increase in the rate of firing. Such a non-linearity is an essential component for a complete representation, since neurons have a maximum rate of firing which cannot possibly be exceeded.

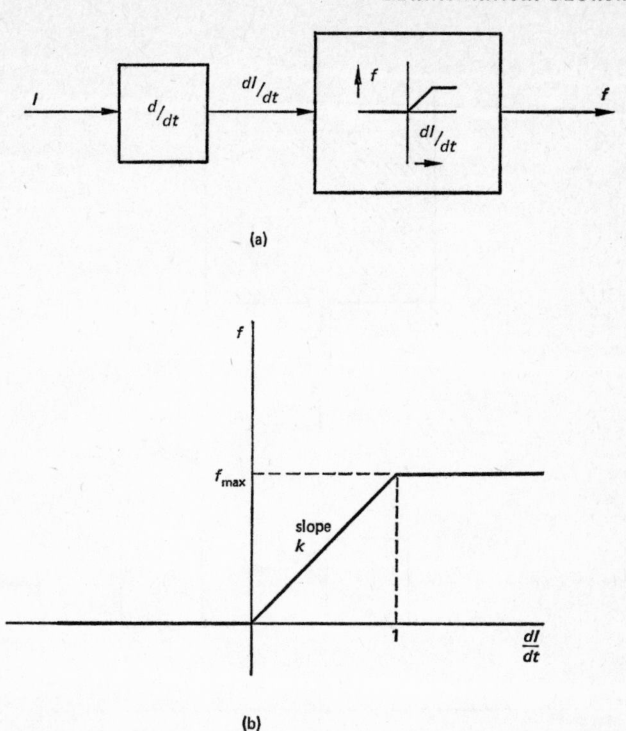

Figure 26 A simple model of a retinal component that is sensitive to increases in illumination. (a) The complete model, (b) the function relating firing rate to the rate of change of illumination.

This non-linearity is included in the model by means of the second block. The input is dI/dt, and a value of f is selected that corresponds to the particular value of dI/dt dictated by the non-linearity, which is exactly as shown in part (b) of the figure.

The output of the overall system is the rate of firing and the system relates this to the input, illumination.

For another example of a mathematical model, suppose, as Figure 27(a) indicates, that the flow out of a fluid container is directly proportional to the volume of fluid in the container. It is desired to construct a mathematical model to show flow of water out and its dependence upon flow in. We need two equations to describe the system. The first relates flow out to volume and is given by

$$\text{flow out} = K \times \text{volume}$$

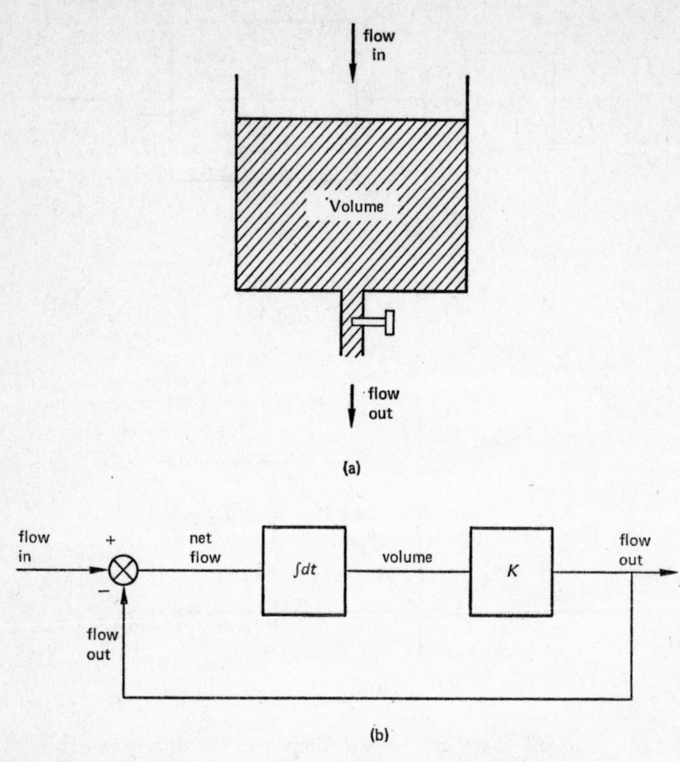

Figure 27 *A fluid container in which flow out is proportional to volume. (a) The physical system. Flow out* $= K_1 \times$ *pressure; pressure* $= K_2 \times$ *volume; and, therefore, Flow out* $= K \times$ *volume; where* K_1, K_2 *and* K *are constants. (b) A model of the system.*

where K is a constant. The second equation relates volume to flow. It was explained in Chapter 2 that volume is the integral of flow. In the present case flow means net flow into the container which is given by flow in minus flow out.

$$\text{volume} = \int_0^t (\text{flow in} - \text{flow out})\, dt$$

It is assumed that the container was empty at time zero.

A block diagram representation is shown in Figure 27(b). A circle with a cross in it always indicates summation or subtraction of signals according to the sign given. In this case flow out is subtracted from flow in and the difference produces net flow.

This net flow is then integrated to give volume of fluid in the container. Multiplying this by K gives flow out, which is then fed back to the beginning. Note the dependence of flow out on flow in via the other terms. Flow in is quite independent of other terms in the system.

What determines K? The answer is the setting of the tap. As the space through which the flow of water must pass is reduced, K gets smaller since any given volume now produces a smaller flow.

Simple as the model is, it none the less can serve to demonstrate the essence of the approach that is being advanced in this book. We build a model on the basis of observable and reproducible results, i.e. that the flow from the container is proportional to the volume of fluid in the container, and that volume is the integral of the net flow into the container. The parameter K refers to the aperture size, a measurable property of the system. We then arrive at a model that represents a theory of how the system works. The next thing to do is to test the adequacy of the model as an explanatory tool. This may be done by putting test signals into the model and comparing its response to that of the real system.

Since flow out is proportional to volume and since volume decreases as a function of flow out, the model predicts that for a sudden load of water (an impulse of flow) at time zero the volume would increase as a step and then decay exponentially. For a step input of flow the volume would increase until a constant value of volume was reached. These are the unambiguous predictions of the model, and if the real system were to do anything other than this, then this indicates an inadequacy in the model.

To summarise, we build a model on the basis of our understanding of the real system. We then test the adequacy of the overall performance of the model against the behaviour of the real system.

So far in this chapter we have met differentiators, integrators, summing points, and linear and non-linear terms which multiply an incoming signal in order to give an output signal. Also, variables sometimes need to be multiplied together or divided, and for an example of where division is used we return to physiology.

The plasma of the blood has a high protein concentration, while the interstitial space with which it is in contact has a low concentration. As a consequence, protein tends to leave the plasma and enter the interstitial space at a rate which is roughly proportional to the difference in concentration between the two compartments (Guyton and Coleman, 1967). A model of this is shown in Figure 28.

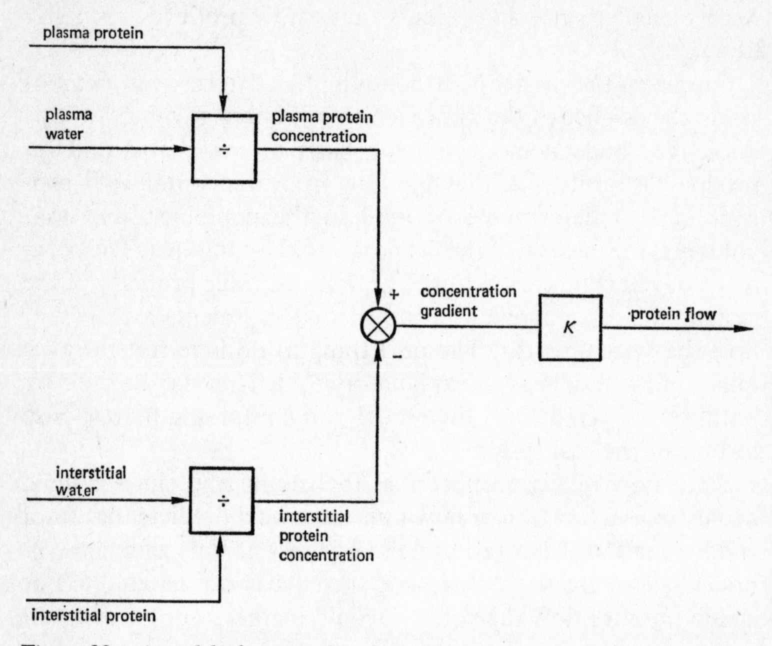

Figure 28 A model of protein flow from the plasma to the interstitial space.

Each protein concentration is given by the respective protein quantity divided by the water content of the compartment concerned. Concentration difference is obtained by subtracting interstitial from plasma concentration. Multiplying concentration difference by the constant K gives rate of movement of protein into the interstitial space. The model of Figure 28 therefore embodies our physiological understanding that flow is proportional to the concentration gradient. K depends upon the size of the barrier for protein exchange.

These simple examples illustrate the use of the block diagram technique as a means of representing a system. However, by now the reader is probably wondering what useful advantages this representation has over simply writing down the equivalent equations in each case. For a complex system such as, for instance, the regulation of blood glucose then a set of, say, forty equations (which may be the minimum possible to do justice to the system) is exceedingly difficult to understand and follow in a logical order. A block diagram incorporating the same information is much easier to follow. The flow of information is clear and somehow intuitively appealing. Parameters are shown as the content of the block, while the variables are shown as arrows.

Laplace Transforms

In the examples of Figures 26 and 27 we encountered signals which were specified as functions of time and operators in the time domain, i.e. integration and differentiation with respect to time. Representation of systems in the time domain describes in a realistic way what is going on, and for this reason is very commonly employed in model-building.

However, Laplace notation is perhaps more commonly used, and after reading Chapter 3 it will be appreciated that there is a direct correspondence between time and Laplace. Models may either be constructed in terms of time-dependent operators and therefore employ time-dependent signals, or in terms of Laplace operators in which case signals must take the form of Laplace transforms. The reader should attempt to be familiar with both forms of representation and how to translate from one to the other. However, once having chosen one form of representation for a particular model, it must be maintained throughout the model. Thus it is meaningless to attempt to use Laplace operators together with time-dependent variables, though this is perhaps the most common mistake made by the beginner.

Laplace notation is at first a little difficult to relate to the real system, but the reason that mathematical models are so often constructed in Laplace terms is that in this way a complex system may often be simplified. In Laplace terms, what was called a signal or time function in the time domain is now the Laplace transform of a signal (i.e. $F(s) = \mathscr{L}f(t)$), while the operators of the system are known as transfer functions. The contents of the block is the transfer function, and this multiplies the input-transform in order to give the output-transform. Figure 29(a)

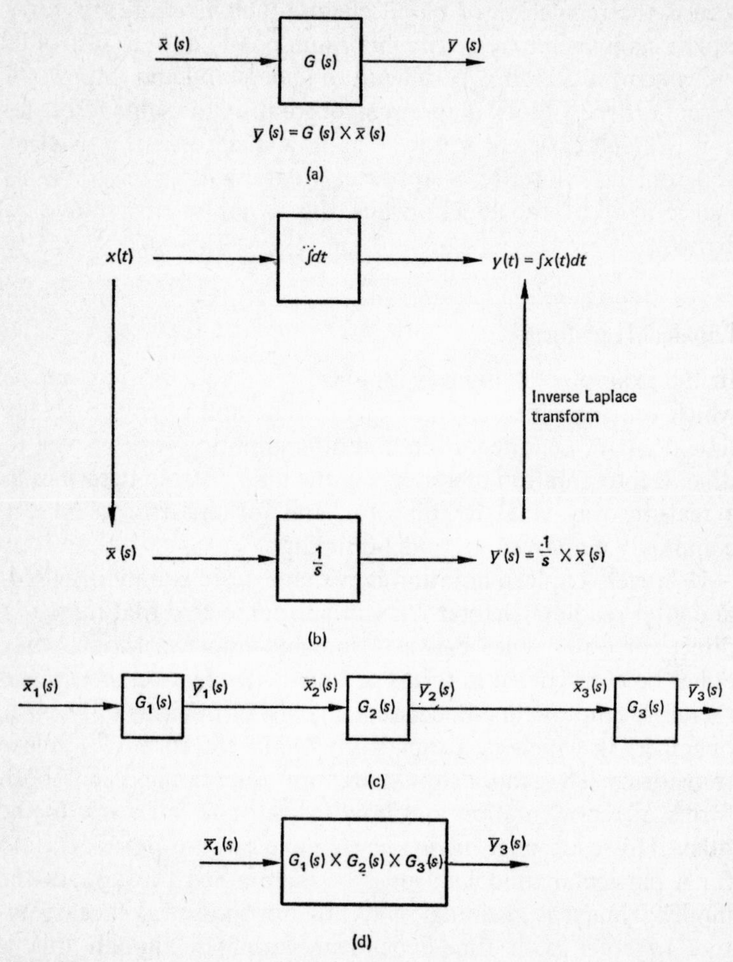

Figure 29 The transfer function. (a) The transfer function G(s) multiplies its input x(s) to give an output y(s). (b) A comparison between time and Laplace. At the top the integrator receives a signal x(t) and gives as an output ∫ x(t) dt. At the bottom the same operation is done in Laplace simply by multiplying by 1/s. In the upper part we do not multiply the input by the contents of the block in order to get the output as we do at the bottom. This means that Laplace represents a simplification. (c) The three transfer functions may be reduced to the single transfer function of (d).

illustrates this, the transfer function being $G(s)$, the s in brackets indicating a Laplace operator. The transform $\bar{x}(s)$ is multiplied by the transfer function $G(s)$ to give the transform $\bar{y}(s)$, alternatively written as $Y(s)$.

$$\bar{y}(s) = G(s) \times \bar{x}(s)$$

The beauty of the transfer function method is that the transfer function always simply multiplies the incoming transform to give the output-transform. Thus a problem set in terms of a differential equation can have an algebraic solution. Multiplication by a transfer function is a considerable simplification over the equivalent operation in the time domain, where no such simple relationship holds except for an operator which is a pure constant. Unfortunately Laplace transforms are only applicable to linear systems.

An example can serve to illustrate the use of the Laplace technique. In Chapter 3 it was explained that if initial conditions are zero, the Laplace transform of the integral of a function of time is given by the Laplace transform of the function, multiplied by $1/s$, i.e.

$$\mathscr{L}[\int f(t)] = [\mathscr{L}f(t)] \times \frac{1}{s}$$

Thus in order to integrate in Laplace terms we simply multiply by $1/s$. In Figure 29(b) this is shown together with the same operation in the time domain. In the top of the figure a function of time $x(t)$ enters an integrator and an output $y(t)$ is produced such that $y(t) = \int x(t)\,dt$. In the lower part of the figure exactly the same operation is carried out in Laplace terms. The operation of integration with respect to time is replaced by the transfer function $1/s$, since these are equivalent. The Laplace transform of $x(t)$ is $\bar{x}(s)$, and $\bar{x}(s)$ is multiplied by $1/s$ to give $\bar{y}(s)$. The Laplace transform of $y(t)$ is $\bar{y}(s)$ and, if $y(t)$ is inverted, it gives the integral of $x(t)$ as a time function.

Figure 29(c) shows several transfer functions connected in series, and from the basic equation of the transfer function which was given, this network can be simplified. The equations of the system are

$$\bar{y}_1(s) = G_1(s)\bar{x}_1(s)$$

$$\bar{y}_2(s) = G_2(s)\bar{x}_2(s)$$

$$\bar{y}_3(s) = G_3(s)\bar{x}_3(s)$$

$$\bar{y}_1(s) = \bar{x}_2(s)$$

$$\bar{y}_2(s) = \bar{x}_3(s)$$

By substitution in these equations we obtain the result that

$$\bar{y}_3(s) = G_1(s) \times G_2(s) \times G_3(s) \times \bar{x}_1(s)$$

Therefore the three transfer functions of Figure 29(c) may be replaced by the single transfer function of Figure 29(d). This is one of the means by which complex systems may be simplified when using transfer functions. Further use of the Laplace transform is best illustrated by means of working through problems, some of which appear now.

Example 8

An operator carries out the operation of integration on its input signal. Calculate the response to a unit step input, by means of the Laplace technique.

First, the input signal must be converted into a Laplace transform. In Chapter 3 the Laplace transform of a unit step input was given as $1/s$. This term must be multiplied by the transfer function corresponding to integration which is $1/s$, and this gives an output-transform of $1/s^2$, as shown in Figure 30(a). The output, obtained as a Laplace transform, must be inverted. From Table 1 in Chapter 3 we find that the inverse transform of $1/s^2$ is the time function t, and so the answer is a unit ramp t.

The reader should verify this by checking that the integral of 1 with respect to time is given by t. If any difficulty is experienced in understanding what is happening, a very simple practical example will illustrate the point.

The input signal may be considered to be flow and the output signal to be volume. The transfer function represents the operation which relates these terms, in real life a bucket, for instance. With respect to the time domain, if a constant flow of

1 unit/time enters a bucket, then the volume in the bucket is given by t units at time t.

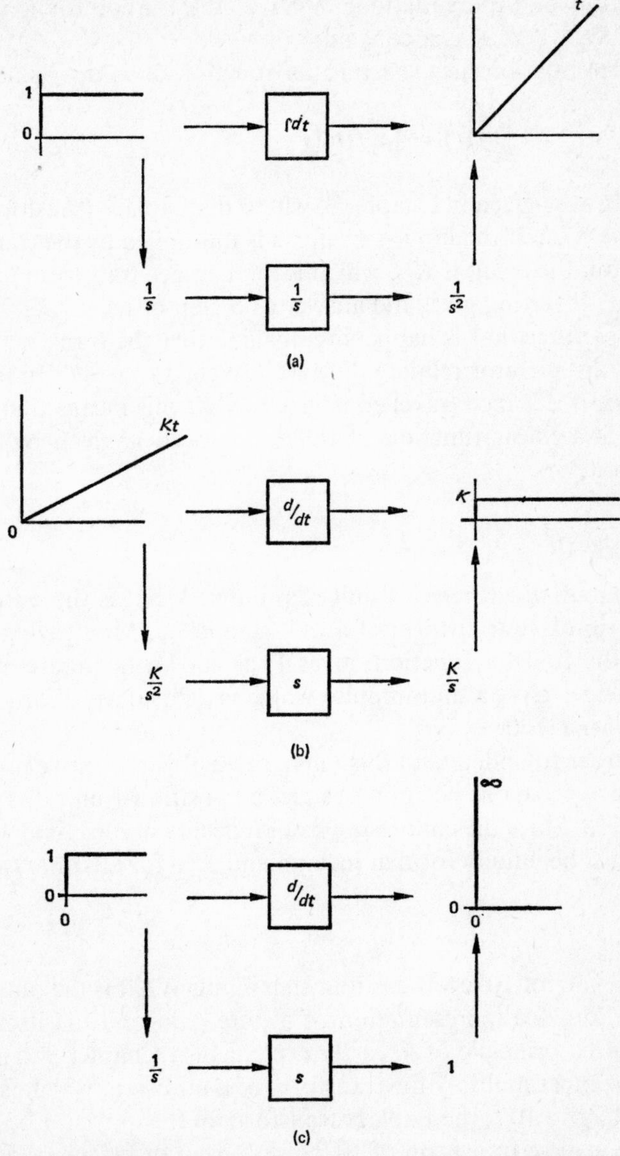

Figure 30 Examples of the use of the transfer function method (for explanation see text).

Example 9

The transfer function shown in Figure 30(b) represents the operation of differentiation. What is the output for a ramp input Kt, where K is a constant?

Differentiation means a transfer function of value s, since

$$\mathscr{L}\frac{d}{dt}f(t) = [\mathscr{L}f(t)]s$$

Details were given in Chapter 3, where the Laplace transform of Kt gave K/s^2. If the input-transform is multiplied by the transfer function, the result is K/s, which is the Laplace transform of the output. Inverting gives the answer as a step of height K.

To picture what is happening imagine that the transfer function is an operator relating distance travelled to speed. In terms of time, if distance travelled is given by Kt this means that distance is a linear function of time and speed is given by K, a constant.

Example 10

A differentiator receives a unit step input. What is the output?

The Laplace transform of a unit step is $1/s$. Multiplying this by s, the transfer function, gives 1, as shown in Figure 30(c). Inverting 1 gives a unit impulse which is defined as ∞ for $t = 0$ and 0 for $t \neq 0$.

In order to understand this, imagine an object to move instantaneously from one position to a second position 1 unit away, at time zero. Since the change in position occurs in zero time, velocity must be infinity for that instant, and zero for all other times.

Example 11

A pure gain of 10 receives a unit step input. What is the output?

The Laplace representation of a pure gain of 10 is itself 10 due to the principle of linearity explained in Chapter 3. Therefore we must multiply the transform of a unit step, which is $1/s$, by 10 to give $10/s$, the Laplace transform of the output. Looking up the inverse transform of $10/s$ gives a step of height 10 as the answer, a not altogether surprising result.

Example 12

Let us now look at a problem from the opposite point of view, and in fact from the viewpoint which the biologist normally has. The biologist applies test signals to a system, observes the responses, and from this information determines the transfer function. The engineer, by contrast, usually has the problem of adjusting the transfer function to give a particular performance.

We are presented with a system and apply a step input to it. As a result we obtain an exponentially delayed response which is shown in Figure 31(a). What is the transfer function of the system?

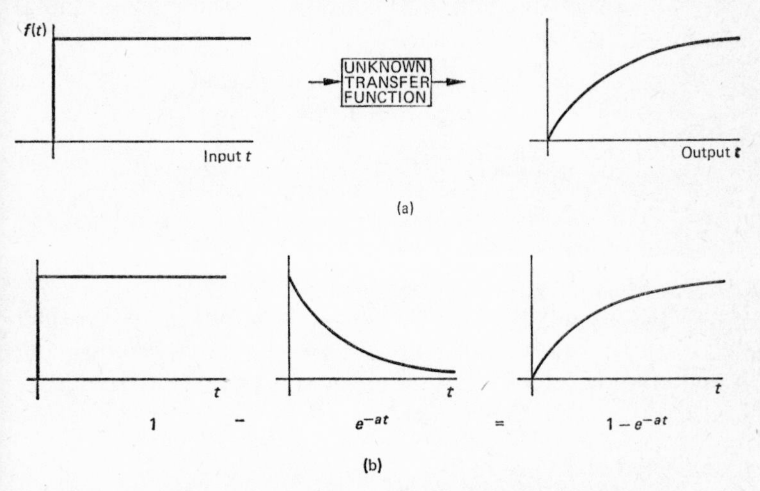

Figure 31 Estimation of an unknown transfer function. (a) For a step input an exponential rise results, (b) the exponential rise is broken into its components: a step and an exponential decay.

An exponentially delayed time function which finally reaches a constant value of 1 is made up of two parts, as indicated in Figure 31(b). The first is a step of height 1 and the second is a decaying exponential having an initial value of 1. The step minus the exponential gives the response shown in Figure 31(a). The output response as a function of time is therefore given as $1-\exp(-at)$, where a is a constant that determines the rate of decay of the exponential. The value of $\exp(-at)$ is 1 when t equals zero, so the response $1-\exp(-at)$ starts from zero.

When t gets very large $\exp(-at)$ approaches zero, so the response tends to 1.

The output can be converted to a Laplace transform.

$$\begin{aligned}\mathscr{L}[1-\exp(-at)] &= \frac{1}{s}-\frac{1}{s+a} \qquad \text{(see Chapter 3)}\\ &= \frac{s+a-s}{s(s+a)}\\ &= \frac{a}{s(s+a)}\end{aligned}$$

The input is a unit step and so it has a Laplace transform of $1/s$. The transfer function $H(s)$ is defined, assuming zero initial conditions, by

$$\begin{aligned}H(s) &= \frac{\text{Laplace transform of output}}{\text{Laplace transform of input}}\\ &= \frac{a/[s(s+a)]}{1/s}\end{aligned}$$

Therefore, $H(s) = \dfrac{a}{s+a}$

In practice we might consider that we need more evidence than this in order to determine the transfer function reliably, and would employ, in addition, ramp or sinusoidal test stimuli.

Exponential and Pure Delays

In the last example a sudden change was made to the input of a transfer function, but although the output started to respond immediately, time was needed before it reached its final value. The output changed exponentially, i.e. the rate of change was proportional to the distance the output was from its final value. An element which produces this output in response to a step input is known as an exponential delay or lag, and is characterised by the transfer function $K/(s+a)$, where K is the gain constant, and a determines the speed of the exponential response. We will have more to say about this transfer function later, since it is very commonly encountered in biological systems. For instance, if the level of illumination is suddenly changed, this demands a change in pupil size. The pupil cannot suddenly reach its new value, but approaches it in an approximately exponential fashion (see also Chapter 7 for a more detailed consideration of the pupil).

Figure 32(a) shows very roughly what the pupil-response to a change in illumination looks like, and serves as an introduction

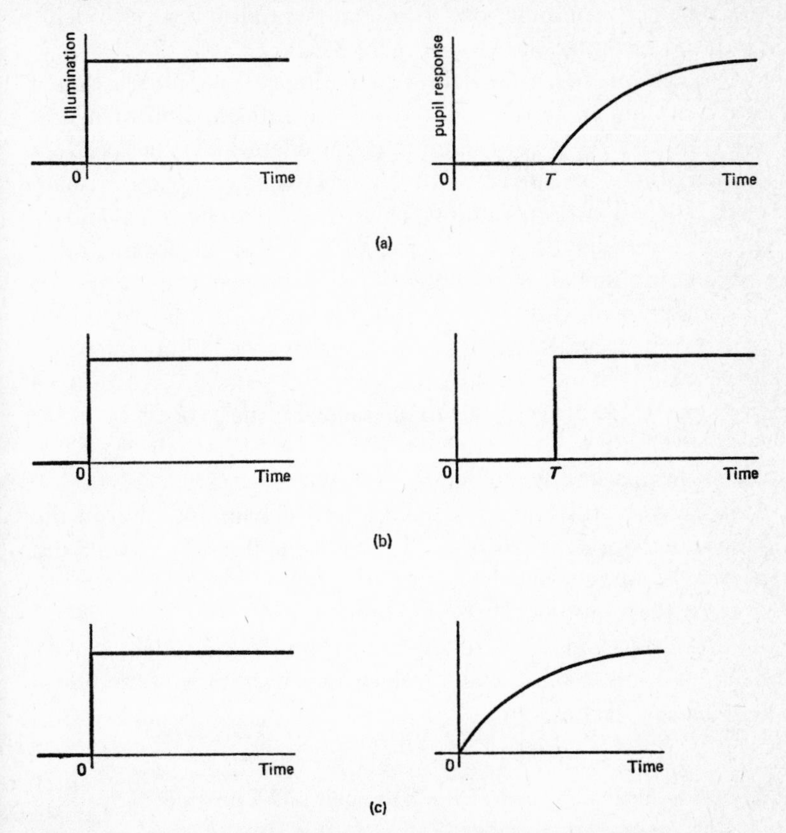

Figure 32 The response of the pupil to a step change in illumination. (a) The response may be seen to be composed of a pure delay of T s *and then an exponentially delayed response. Taking the response apart we have, (b) a pure delay and (c) an exponential lag.*

to another component. At time zero the illumination changes, but a time T elapses before the pupil even starts to move. At time T, the pupil begins its exponential response, T being of the order of 0·16 s (Stark and Sherman, 1957).

Figure 32(b) shows the effect of a pure delay. The signal enters the component that imposes the delay, and T seconds later it appears at the output. Note that apart from being delayed, the signal is not changed by the delay, exactly the same signal comes out as goes in except that it is shifted in time. By contrast, as Figure 32(c) shows, the exponential delay changes the shape of the signal. A combination of an exponential and a pure delay gives the response shown in Figure 32(a).

One reason for pure delays in biological systems is that a finite amount of time is needed for the transmission of neural information. As a very minimum consideration, the speed of conduction by neurons is relatively slow when compared with an electronic network or radio transmitter, where the signal travels at an exceedingly fast speed (for all practical purposes, except space communication, in zero time). Each synapse delays the transmission of information by between 0·3 millisecond to several milliseconds, while the conduction speed along the nerve fibre varies between 1 and 100 metres per second, according to the type of fibres. In order to characterise the process of nerve conduction we would have to include pure delays if the accuracy of the simulation demanded it.

In quasi-behavioural systems such as accommodation in the eye, and more so in tasks which require skill, such as tracking a moving target, the delay in central processing is more important than peripheral nerve conduction.

This discussion of the fundamental components enables us to begin to analyse some biological systems by means of the transfer function technique.

A Mathematical Model of the Stomach and Intestine

The object of this section is to construct a highly simplified model which represents the transport of water in the stomach and intestine.

Water taken by mouth first passes along the oesophagus and then enters the stomach. For experimental purposes water may be placed in the stomach by means of a tube. Water leaves the stomach at the lower end by the pyloric sphincter and enters the intestine. After reaching the intestine it is able to cross the gut-wall and enter the blood. A model of what happens is shown in Figure 33(a) and (b).

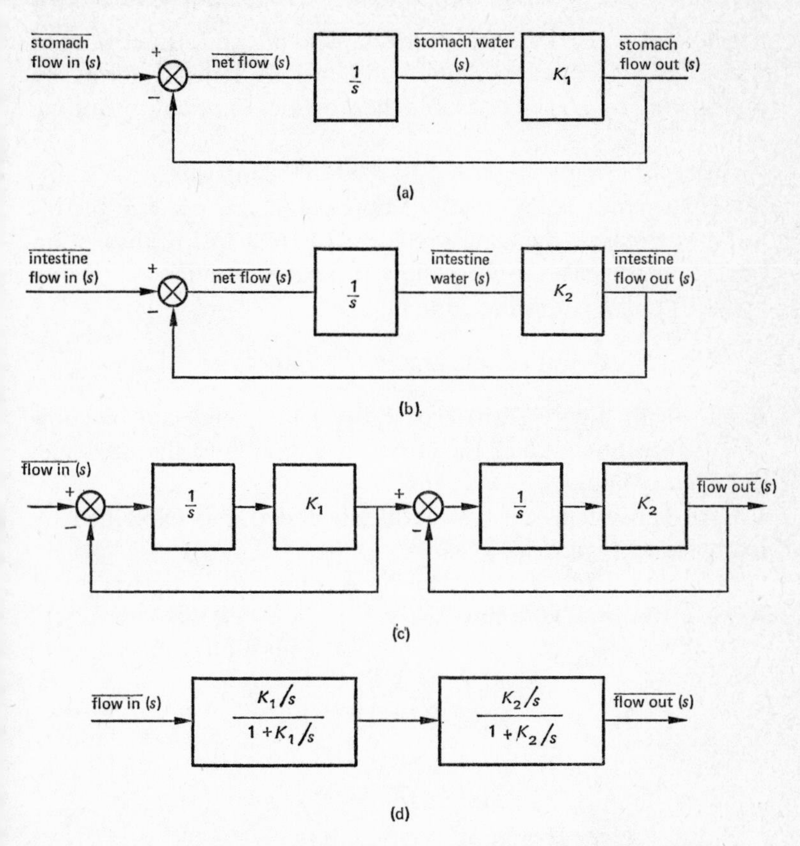

Figure 33 A mathematical model of the gut transport of water. (a) The stomach, (b) the intestine, (c) the stomach and intestine in combination, (d) reduction of the stomach–intestine model to a simpler expression.

O'Kelly, Falk and Flint (1958), and Evans (1949), showed that if a load of water is suddenly placed in the stomach by means of a tube, the volume of water in the stomach decreases exponentially. This implies that flow out of the stomach is proportional to stomach contents, a common situation and in fact the one which we analysed in Figure 27 of this chapter. We can say that

$$\text{flow out of stomach} = K_1 \times \text{stomach volume}$$

where K_1 is a constant. This relationship enables the model of Figure 33(a) to be drawn. Flow out is made proportional to stomach volume, via K_1, and is subtracted from flow in to give net flow of water into the stomach. The integral of net flow of water gives stomach volume (obtained in Laplace terms by multiplying by $1/s$). We assume the stomach to be empty at time zero.

A similar argument may be applied to the intestine (for details see Toates and Oatley, 1970). As the volume of water in the intestine increases, then the surface area over which absorption can occur increases, and so flow from the intestine is roughly proportional to intestine contents, i.e.

$$\text{flow out of intestine} = K_2 \times \text{intestine volume}$$

as shown in Figure 33(b). These two subsystems can be combined since flow out of the stomach is flow into the intestine. See Figure 33(c).

Each subsystem must be reduced, and this is explained by reference to Figure 34.

In Figure 34(a) $\bar{y}(s)$ is given by $\bar{e}(s)$ multiplied by $G(s)$, and $\bar{e}(s)$ is given by $\bar{x}(s)$ minus $\bar{y}(s)$.

$$\bar{e}(s) = \bar{x}(s) - \bar{y}(s)$$

$$\bar{y}(s) = G(s) \times \bar{e}(s)$$

therefore

$$\bar{y}(s) = G(s) \times [\bar{x}(s) - \bar{y}(s)]$$

$$\bar{y}(s) = G(s) \times \bar{x}(s) - G(s) \times \bar{y}(s)$$

$$\bar{y}(s) \times [1 + G(s)] = G(s) \times \bar{x}(s)$$

$$\bar{y}(s) = \frac{G(s)}{1 + G(s)} \times \bar{x}(s)$$

The transfer function relating $\bar{y}(s)$ to $\bar{x}(s)$ is given by $G(s)/[1 + G(s)]$ and the system of Figure 34(a) may be replaced by the single transfer function of Figure 34(b). In comparing Figures 33 and 34 we can reduce Figure 33(c) by noting that K_1/s and K_2/s in Figure 33(c) correspond to $G(s)$ in Figure 34.

Figure 33(d) can be drawn in place of Figure 33(c), and the transfer function relating flow from intestine to flow into stomach is given by

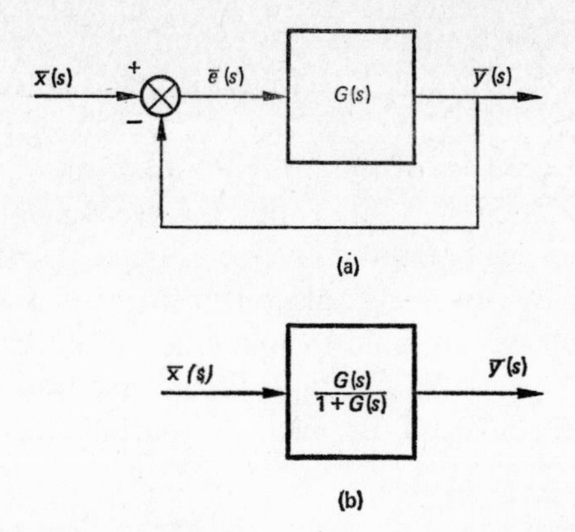

Figure 34 Reduction of a feedback loop. (a) The system with a unity feedback loop around it, (b) the system in its simplified form.

$$\frac{K_1/s}{1+K_1/s} \times \frac{K_2/s}{1+K_2/s}$$

which may be simplified to give

$$\frac{K_1}{s+K_1} \times \frac{K_2}{s+K_2}$$

The system is now constructed and so the next step is to test its predictions. Let us find the response of the system to a sudden stomach load of water. If we assume that 1 unit of water is applied in a very short period of time, then input flow corresponds to a unit impulse. Obviously this cannot be exact, but it is sufficiently true to be employed as a first approximation.

$$\overline{\text{flow out}}\,(s) = \begin{bmatrix}\text{transfer function}\\ \text{between flow in}\\ \text{and flow out}\end{bmatrix} \times [\overline{\text{flow in}}\,(s)]$$

$$\overline{\text{flow out}}\,(s) = \frac{K_1}{s+K_1} \times \frac{K_2}{s+K_2} \times [\overline{\text{flow in}}\,(s)]$$

flow in $=$ unit impulse $=$ 1 as a Laplace transform

$$\overline{\text{flow out}}\,(s) = \frac{K_1}{s+K_1} \times \frac{K_2}{s+K_2} \times 1$$

In order to invert this term we must split it up into its component parts. We do this by means of the partial fraction method, which was explained in Chapter 3. A very common mistake is to assume that since the term we wish to invert is made up of the product of two simpler terms, we merely invert each of them and the answer is the product of the inverse transform of each component. This is incorrect since the inverse Laplace transform $F_1(s)F_2(s)$ does not equal the inverse Laplace transform of $F_1(s)$ multiplied by the inverse Laplace transform of $F_2(s)$. The relationship which we are able to make use of is that the inverse transform of $[F_1(s)+F_2(s)]$ equals the inverse transform of $F_1(s)$ plus the inverse of $F_2(s)$. By means of partial fractions

$$\begin{aligned}
\frac{K_1K_2}{(s+K_1)(s+K_2)} &= \frac{A}{s+K_1}+\frac{B}{s+K_2} \\
K_1K_2 &= A(s+K_2)+B(s+K_1) \\
K_1K_2 &= As+AK_2+Bs+BK_1 \\
0 &= A+B \\
K_1K_2 &= AK_2+BK_1 \\
A &= -B \\
K_1K_2 &= -BK_2+BK_1 \\
K_1K_2 &= B(K_1-K_2) \\
B &= \frac{K_1K_2}{K_1-K_2} \\
A &= -\frac{(K_1K_2)}{(K_1-K_2)} \\
\overline{\text{output}}\ (s) &= \frac{K_1K_2}{K_1-K_2}\times\left[\frac{1}{s+K_2}-\frac{1}{s+K_1}\right]
\end{aligned}$$

We can now invert this by using the fact, given in Chapter 3, that $\mathscr{L}K\exp(-at) = K/(s+a)$.

$$\text{output}\ (t) = \frac{K_1K_2}{K_1-K_2}\times[\exp(-K_2t)-\exp(-K_1t)]$$

Therefore the flow out of the intestine is described by two decaying exponentials. At time zero, both terms are equal and opposite and so the response then is zero. At time infinity both terms are zero, so the flow rate is zero at that time. However the positive exponential will always decay more slowly than the negative exponential, whatever values are chosen for K_1 and K_2, since if K_1 is bigger than K_2 then $(K_1 - K_2)$ is negative. This changes the sign back to positive again, and so the flow from the intestine rises and then decays, as shown in Figure 35.

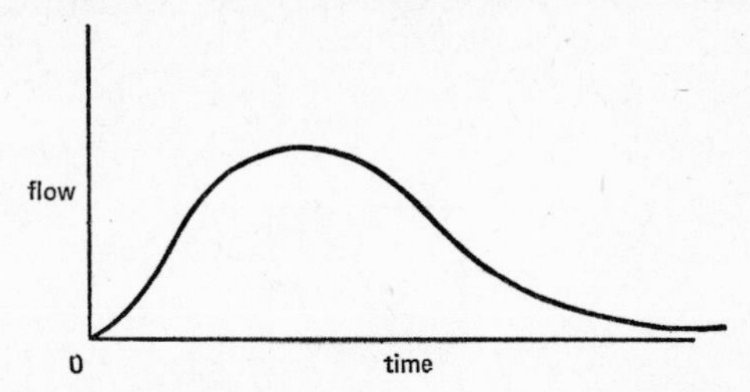

Figure 35 The response of the model.

What happens to the volume of water in the intestine? Intestinal volume was multiplied by K_2 to give flow out of the intestine, therefore in order to get intestine water we must divide flow by K_2.

$$\text{intestine water }(t) = \frac{K_1}{(K_1/K_2) - 1} \times [\exp(-K_2 t) - \exp(-K_1 t)]$$

The volume of water in the intestine also rises and decays after a sudden load of water in the stomach, in the fashion shown in Figure 35. The model enables the dynamics of the intestine to be understood in terms of the component operations involved, i.e. the stomach to intestine and intestine to blood dynamics.

A Simple Model of Drinking

Let us suppose, for the purpose of constructing a crude model, that drinking rate is directly proportional to the deficit in body water (this is approximately true for the rat. See Stellar and Hill (1953) for details). What is the drinking response to a steady, continuous flow of water at rate N, out of the body fluid pool?

Figure 36(a) shows the transfer function representation of the system. Drinking is proportional to deficit in body water, the constant of proportionality being given by K. Rate of removal of fluid forms the input to the system, and the integral of this is the deficit. Drinking must also be an input to this integrator but with an opposite sign to that of the rate of removal.

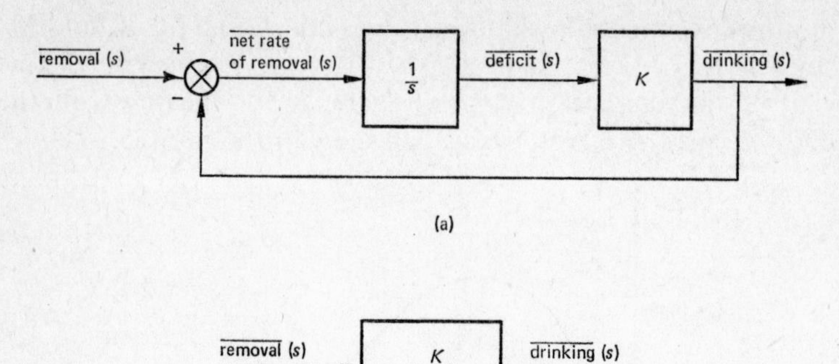

Figure 36 *A model of drinking. (a) The basic model, (b) the reduced model.*

$$\text{deficit} = \int_0^t (\text{removal} - \text{drinking})\, dt$$

We saw in the last section that a system of the kind shown in Figure 36(a) may be reduced to that of Figure 36(b). The input is a step of height N which has a Laplace transform of N/s.

$$\overline{\text{output}}\,(s) = (\text{transfer function}) \times [\overline{\text{input}}\,(s)]$$

$$\overline{\text{output}}\,(s) = \frac{K}{s+K} \times \frac{N}{s}$$

Again we use partial fractions

$$\frac{KN}{s(s+K)} = \frac{A}{s+K} + \frac{B}{s}$$

$$KN = As + B(s+K)$$

$$KN = As + Bs + BK$$

$$BK = KN$$

$$B = N$$

$$A = -N$$

$$\overline{\text{output}}\,(s) = \frac{N}{s} - \frac{N}{s+k}$$

$$\text{output}\,(t) = N - N \exp(-Kt)$$

The first term is a step of height N. The second is a decaying exponential which has the value N at time zero. Therefore at time zero these two terms cancel out so that drinking starts off with a value of zero. For large values of time the exponential has essentially decayed to zero so that drinking is at rate N. This means that in the steady-state, drinking rate is exactly equal to rate of removal of fluid.

What happens to the deficit? It may be seen from Figure 36(a) that drinking equals deficit multiplied by K. Therefore,

$$\text{deficit} = (\text{drinking rate})/K$$

Deficit finally reaches a constant value of N/K. Increasing the term K decreases the deficit for any given rate of removal.

What is the drinking response for a unit impulse input?

$$\overline{\text{output}}\,(s) = \frac{K}{s+K}\,[\overline{\text{input}}\,(s)]$$

$$\text{unit impulse} = 1$$

$$\overline{\text{output}}\,(s) = \frac{K}{s+K}$$

$$\text{output}\,(t) = K\exp(-Kt)$$

It is a decaying exponential having an initial value K.

The mathematically keen reader may be interested to see how much water is drunk in response to the unit impulse.

$$\begin{aligned} \text{water drunk} &= \int_0^\infty (\text{drinking rate})\,dt \\ &= \int_0^\infty K\exp(-Kt)\,dt \\ &= K\times[(-1/K)\exp(-Kt)]_0^\infty \\ &= 1 \end{aligned}$$

One unit of water is drunk, which is exactly the same as the amount of water removed since the area under a unit impulse is 1.

Linearity and Non-Linearity

Our work is made relatively easy by the transfer function method. From the input–output relationship of a system, using

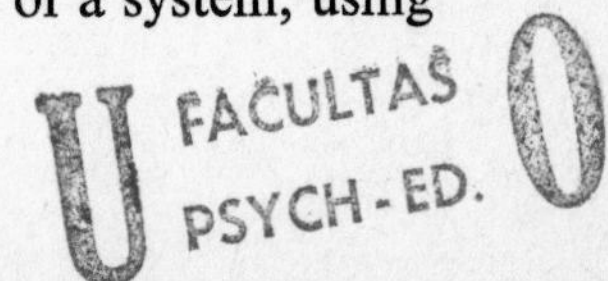

step, ramp and sinusoidal stimuli (the latter are discussed in Chapter 7), we can obtain the transfer function which characterises the system. Complex systems may often be reduced to something simpler, and on the basis of the step and ramp responses we can predict the response to a new untested stimulus.

Having said this, it must be pointed out regretfully that life is decidedly non-linear, and the number of problems which can be fully analysed by linear techniques such as the Laplace method is limited. However, the student ought not to feel too discouraged by realising the limitation of the linear methods which he or she is patiently learning. It is sometimes remarkable just how far one can get by using these techniques. Even if a thorough analysis of the kind we have carried out for linear systems is precluded, we may still be able to extract certain very important features of the system.

Several means are available for handling non-linear systems. As the simplest approach, linearity may be assumed and normal linear techniques used. For instance test signals may be applied to a system, the response measured, the linear fit estimated, and from this information a transfer function obtained. As an approximation, the transfer function may be able to characterise the essential nature of the system and therefore provide useful information. Alternatively, though, we are in danger of missing the most interesting features of the system if we endeavour at all costs to find a linear transfer function. Without some knowledge of the system it is difficult to provide hard and fast rules for the validity of linear techniques.

Some systems behave pseudo-linearly provided that only small signals are employed. For instance, Figure 37 shows a non-linear component which is almost linear over the range indicated, having a slope of R. If we ensure that all signals are confined to such a range, then the use of linear techniques is valid. However, it cannot be over-emphasised that any biological system will behave non-linearly if sufficiently large signals are employed.

For some systems the departure from linearity is so severe that to assume linearity is at best, valueless, and at worst, positively deceptive, so there is no alternative to a rigorous non-linear analysis. Sophisticated mathematical methods are avail-

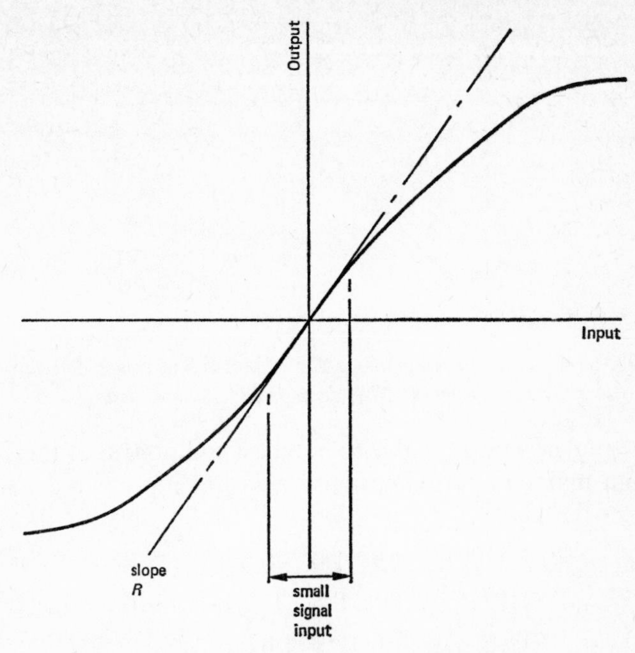

Figure 37 *A non-linear relationship between two terms. Over a limited range we might not do too much injustice to assume that the system is linear with a slope of R.*

able for the analysis of non-linear systems, but they are beyond the scope of an introductory text.

Although analysing real systems by means of the stimulus–response relationship is difficult where non-linearities are involved, representation of known or estimated non-linearities in a model of a system presents few problems. The tool which enables us to handle these is the computer. The essence of the method, described later, is that one makes the computer behave in the same way as the system we wish to study.

Consider, for example, the problem of simulating the non-linear component shown in Figure 38. It consists of two regions, each of which require a separate equation for their description.

1. Input below 1·0. The output equals input times 1·02.
2. Input above 1·0. The output equals 1·02.

Clearly if we are interested in input signals which venture to a value greater than 1·0 it is nonsense to attempt a linear approach.

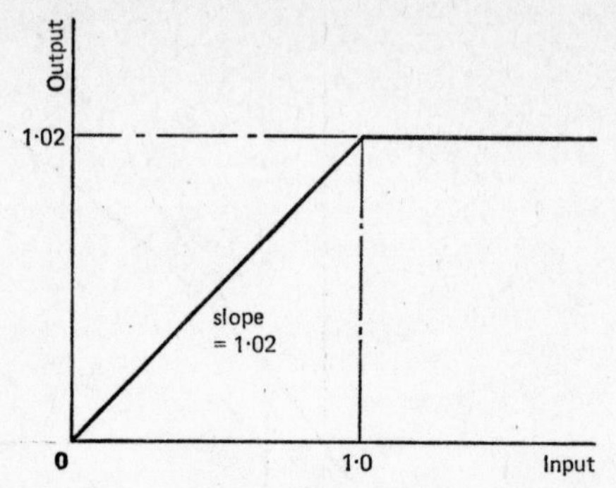

Figure 38 A non-linear component. The system is linear provided that the input is less than 1·0, *otherwise it is most decidedly non-linear*

The digital computer is able to handle a component of this form without making approximations. We can simply make a statement of the kind:

```
'IF' INPUT 'LE' 1·0
'THEN' OUTPUT := INPUT × 1·02
'ELSE' OUTPUT := 1·02;
```

which reads: 'Make a variable called output equal to 1·02 times the input, if the input is less than or equal to 1·0, otherwise make the output equal to 1·02.' A similar operation may be carried out on the analogue computer.

The significance of the position that a non-linear component occupies in a system is shown by the following example, which emphasises the importance of considering non-linearity.

First we examine the linear system of Figure 39(a). A signal is multiplied by 10 and then differentiated so that the output is $(d/dt)10x$. In Figure 39(b) the order of the components is reversed and we now obtain $10(d/dt)x$ which is exactly the same as $(d/dt)10x$. For linear components the order is unimportant.

Contrast this result with that shown in Figure 39(c) and (d) for a non-linear system. In Figure 39(c) a signal x forms the input to a non-linear operator which shows saturation characteristics and gives a maximum output of 1, which is reached at an input of 1. Let us give x the value 1, thus making the output, of the non-linearity, 1. This means that the input to the linear term is 1, giving an output y of 100 units.

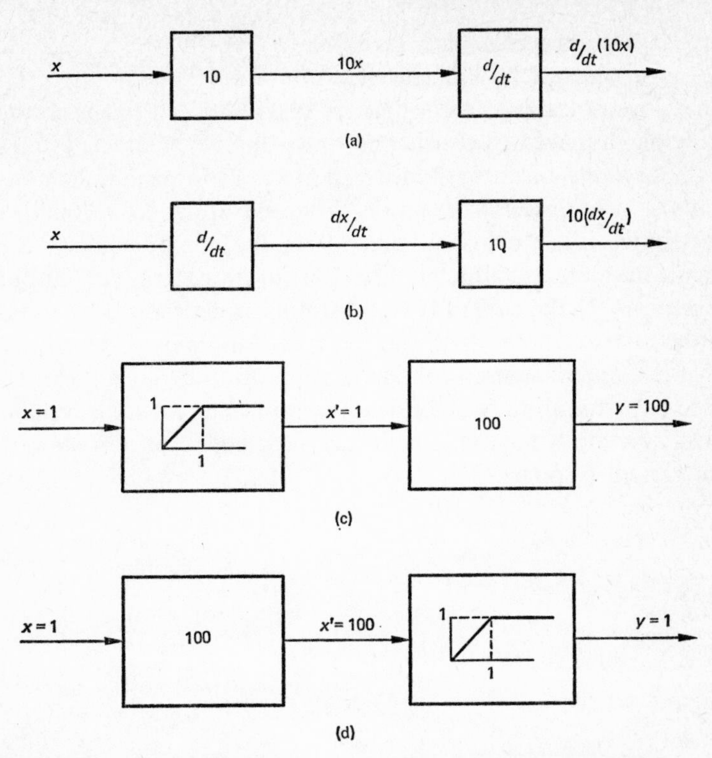

Figure 39 Linear and non-linear systems. The order of components in (a) may be changed to that of (b) without affecting the input–output performance of the overall system. Attempting to do the same thing with the system of (c) in order to give (d) produces a drastic effect on the output, as the numbers in the diagram show.

In Figure 39(d) the effect of changing the order of the components may be seen. The input x is still 1. It is now multiplied by 100 to give an input of 100 units to the non-linear component. An input of 100 can only give an output of 1, since beyond the saturation point, however much the input is increased, the output will remain at 1·0. In this case y equals 1, as opposed to 100 in Figure 39(c), demonstrating the dramatic effect of changing the order of the components. As a general rule, for non-linear components the order is important but for linear systems it is always unimportant.

In Chapter 2, the principle of superposition was discussed. It will be recalled that if an input x_1 gives a response y_1 and an input x_2 a response y_2, then an input $x_1 + x_2$ will give a response $y_1 + y_2$ if the system is linear. For non-linear systems the principle does not apply. Although, as an example to illustrate the principle, a mass was considered, the principle of superposition applies to any linear system irrespective of how many components it contains. For example, it would apply to a complex electronic circuit provided that all of the components of the circuit had linear characteristics. For linear systems the output in response to the sum of two test stimuli will always be the sum of the response to each stimulus applied on its own. This principle is commonly used for the simplification of linear systems since two stimuli may enter at different points into the system, but again the behaviour is entirely predictable as the sum of component responses.

5 FEEDBACK CONTROL SYSTEMS

Feedback in Engineering and Physiology

On a cold morning a man switches on an electric heater in his living room and then leaves home for the day. In applying heat to the room he hopes that a certain temperature will be achieved. This simple example enables two concepts in control engineering to be introduced: the desired value of a control system and its actual value. There is a temperature that the man hopes will be obtained, let us say 18 °C, and this we may call the desired value. Desired value is also known as the input to the system. The actual temperature of the room forms what is known as the output or the actual value of the system. The man, heater and room may be looked upon as forming a system whose object is to make the room temperature's actual value equal to the desired value of 18 °C. How near the actual value comes to matching the desired value is a measure of the success of the control action.

A diagram can serve to illustrate the flow of information in the system – in the present example it does no more than that – and is not the same as a formal block diagram. As Figure 40(a) shows, the man estimates, on the basis of the temperature required and his experience of heating the room, how many heaters or bars of a heater to switch on. The relationship between the man's notion of the temperature he wants and the amount of heat switched on is marked 'man' in the diagram, meaning that the man determines the input–output relationship of the first component in the system. Let us say he switches on two bars of an electric heater. The electric heater determines the relationship between the man's command and the rate of heat production, as represented by the block marked 'electric heater'. For example, the relationship is such that two bars give a rate of

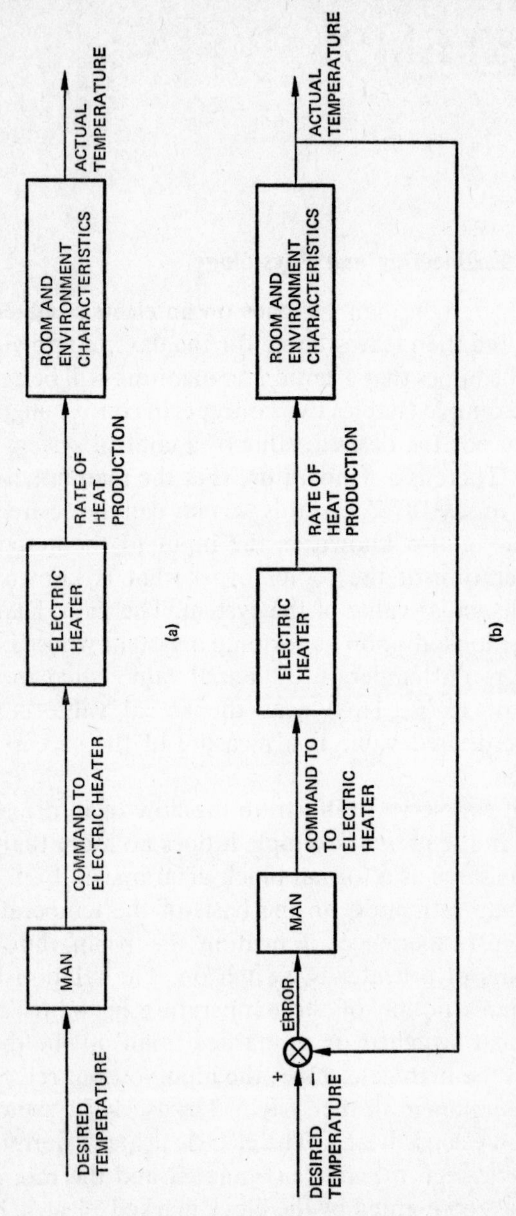

Figure 40 *Room heating systems. (a) Open loop, (b) closed loop.*

heat production of 2 kW. The block marked 'room and environment characteristics' fixes the relationship between rate of heat production and the actual temperature of the room. Thus, if the room is well insulated, a relatively high temperature is produced for a given rate of heat production. At this point we have completed the system, since we have a temperature output for a temperature input.

How well could we expect the system shown in Figure 40(a) to work? It might be expected to work satisfactorily provided that the characteristics of the three components in the loop remain constant, and provided that no unforeseen disturbances appear. We can expect constancy from the block marked 'electric heater', since (electricity board permitting) we will always obtain 2 kW for two switches in the 'on' position. Similarly, we will assume that our human operator has fixed characteristics in that he always puts two switches in the 'on' position. It is in relation to the final component in the loop, 'room and environment characteristics', that there is likely to be trouble. Let us suppose that the weather conditions have been constant, and that on returning home in the evening the room has been at about the desired temperature. Then one day the weather suddenly changes and the outside temperature falls. Due to increased heat loss, the room temperature which results is lower for the same rate of heat production. However the system is quite incapable of doing anything about this, since it is unaware of the actual temperature. Conversely, if the outside temperature became warm the electric heaters might not be necessary, but they would continue to produce heat at the same rate, possibly taking the temperature far above what is needed. The essential point which this example illustrates is that the overall performance of a system of the kind shown in Figure 40(a) is highly sensitive to an external disturbance. It is also highly sensitive to changes in the operating characteristics of any component in the system. If there is a voltage reduction then this changes the characteristic, 'electric heater', and has a profound influence on the output.

We will now consider what would happen if our subject stayed at home all day. If the weather were to turn cold he would switch on additional heating, but if the sun were to shine he would turn it off. Essentially what he is doing is to make a comparison between the actual room temperature and the temperature that he desires, and to take corrective action dependent upon this comparison. Figure 40(b) shows a representation of the flow of information in this system. Actual temperature is fed back and subtracted from desired temperature in order to give the error (in the context of feedback control the desired value is also known as the reference value and, in some situations, as the set-point). The circle with a cross in it represents an error detector. The feedback loop comprises feedback of information, i.e. the temperature sensation that is compared with the man's standard. Some students become worried when no hardware corresponding to the feedback loop can be found; lines on the diagram are not necessarily comparable to electric wires.

If actual temperature equals desired temperature then the signal produced at the error detector is zero, and presumably the man takes zero action, i.e. he does nothing to change the state of things. If there is a positive error signal this means that actual temperature is below desired temperature, and the error is the cue for the man to increase the heating. A negative error signal is the result of actual temperature being greater than desired temperature, and leads to the heating being turned down. Thus error is a factor of immense significance in negative-feedback control systems since the action taken is on the basis of the error.

Figure 40(b) represents what is known as a closed-loop control system, or more specifically a negative-feedback control system. Output is subtracted (hence *negative* feedback) from input and the control action is dependent upon this comparison. In such systems we have a constant check that the output is at the value that we desire. By contrast, the system shown in Figure 40(a) is open-loop in that there is no observation of the output, and consequently the control action is quite independent of the output. As Wilkins (1966) points out, the expression 'open-loop' is somewhat unfortunate, since an open loop is a contradiction in terms. There is no loop at all in an open-loop system, merely a one-way flow of information. However, the term is firmly established in the vocabulary of engineering, so there is little we can do about it. Whereas in the open-loop situation the output value is critically dependent upon both disturbances entering the system and the characteristics of the loop components, a closed-loop system is relatively insensitive to changes in its components and can compensate for disturbances. As we have seen, the environment may change drastically, but in the closed-loop system each change is the stimulus for a correcting action.

It is of course possible to control accurately the temperature of a room without its inhabitant having to stay at home and switch electric heaters up and down. The thermostat is a device for sensing temperature and actuating heating mechanisms in response to a drop in room temperature below a specified set-point. However it performs essentially the same task as the human error detector.

In order to illustrate a fundamental problem encountered in feedback control, we can discard the electric heater and instead heat the room by a radiator. The radiator is switched on automatically when the temperature falls below a particular set-point, switching being under the control of a thermostat.

At the point in time when we start our observations the temperature is below the set-point, and therefore the radiator is switched on. The temperature rises until the set-point is reached and then the flow of water to the radiator is switched off. However, though the flow of water is zero the radiator is of course still hot and is still giving heat to the room, consequently the room temperature rises above the set-point. The system is powerless to do anything about this, despite the appearance of a new error. Gradually heat loss from the room will bring the temperature down again and when it is below the set-point the radiator will be switched on and the cycle repeated. The result is an oscillation between an upper and a lower limit – a phenomenon known as hunting. The output appears to hunt for the input, and this is a characteristic also exhibited by biological control systems under some circumstances.

In my experience, one of the most common sources of misunderstanding that the student makes in connection with control-systems terminology is that if he or she is asked what is the input to the system, the answer is given that 'the rate of heat production is'. Although the rate of heat production is an essential component between input and output, it is not what we call the input to the system. It is easy to confuse the source of energy supplied to the output with the source of information that serves to start the whole process going, i.e. the desired value of 18 °C. The latter is the input.

HOMEOSTASIS

Just as the man's external environment is controlled by the regulation of room temperature, then so his internal temperature is maintained near to a norm or set-point. The healthy functioning of the body is critically dependent upon the fact that body temperature, blood volume, sodium concentration, and many other quantities do not depart far from fixed set-points. *Homeostasis* is the name given to the activity whereby the important functions of the body are maintained relatively constant in the face of disturbances. Regulation is brought about by negative-feedback loops, so that in principle the control of body temperature may be described in the same terms as the control of room temperature by means of a thermostat. The fact that the physical hardware is quite different may for the purposes of explanation and description of performance be relatively unimportant.

Like their equivalents in engineering, homeostatic controllers are equipped with sensors that note any departure from normal in the controlled quantity, and effectors that take corrective action in response to such an error. As an example of a homeostatic system we will discuss the control of body temperature.

Body Temperature Regulation The body temperature of a healthy subject is maintained at approximately 37 °C, and only rarely does it depart far from this. In the case of an animal such as the cat, for a change in environmental temperature of 35°, body temperature changes by only about 1°. By contrast, for the lizard, a cold-blooded animal and one therefore not in possession of a temperature control system, the same environmental change causes a 35° change in body temperature (Gordon, 1972). Whereas the lizard is at the mercy of its environment, the human body is equipped with reflexes which take corrective action when a challenge is encountered. Of course both man and lizard may be able to move to a new environment – an example of behavioural as opposed to reflex homeostasis.

When body temperature rises above normal (or in control theory language, output exceeds the reference value) the human body takes the following corrective actions:

1. The sweat glands pour sweat on to the skin, which by evaporative heat loss cools the body.
2. The blood vessels taking blood to the skin dilate, thus increasing heat loss through the skin.

When body temperature falls below the set-point the following corrective actions are set in motion:

1. The blood vessels of the skin constrict, which has the effect of lowering the blood flow near the skin and consequently reducing heat loss by this route.
2. General metabolism is increased.
3. The muscles are caused to oscillate, a phenomenon that we are familiar with as shivering, which generates heat.

These mechanisms are summarised in Figure 41. The responses are mediated via the autonomic nervous system, and occur

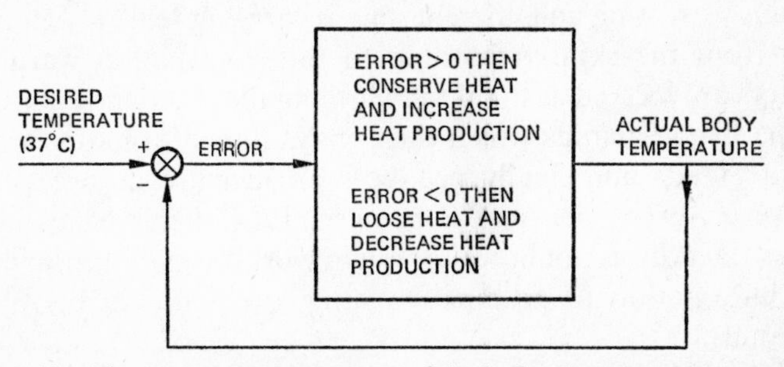

Figure 41 The temperature responses of the body. A positive error signal means that actual temperature is below the set-point. Heat conservation and an increase in heat production are called for. A negative error signal means that the actual body temperature is above the set-point. This is the stimulus to lose heat and to decrease heat production.

rapidly following a slight disturbance in body temperature. Temperature sensors note any departure from normal, and a neural connection exists between the sensor and effector mechanisms. In the brain (to be specific, in the hypothalamus) temperature-sensitive neurons are to be found. They are ideally located for the detection of central temperature because, being in intimate contact with the blood, their activity is able to reflect any disturbance. Of course, temperature-sensitive neurons that transduce temperature into a neural signal are not enough to account for the control action. The reference value must also have some neurological embodiment, and this aspect of biological controllers is perhaps the least well understood. If the rate of firing in the temperature-sensitive neuron is to be compared with some standard, then this standard must take the form of a neuron that has a constant rate of firing unaffected by temperature. There do indeed appear to be both temperature responsive and temperature unresponsive neurons in the preoptic region (Nakayama *et al.* 1963). These neurone circuits are comparable to the thermostat that controls room temperature. When blood temperature rises above normal, the heat loss mechanisms are activated. In humans, at least, metabolic heat generation appears to be more closely dependent upon skin receptors (Benzinger, 1964, 1969).

Originally it was assumed that the anterior hypothalamus is the site of a synaptic connection between efferent nerves that produce sweating and afferent nerves carrying sensory information from the skin. According to this explanation warm receptors are located at the skin; warming the skin sends impulses to the hypothalamus which then relays this information to the sweat glands and finally produces an increase in sweat rate. However, experiments of Benzinger (1964, 1969) have shown that in fact this is not how the system works. It now appears that the information to produce sweating originates in the hypothalamus.

If sweating were to be caused by warm receptors at the skin, in a warm environment we would expect to find some kind of correlation between skin temperature and the rate of sweating. However, following exercise, sweating is accompanied by a decrease in skin temperature. As Benzinger points out, this must be the case since the cooling effect of sweating forms the basis of the thermoregulatory mechanism of cooling. An increase in cranial internal temperature produces an increase in sweating which in turn causes a decrease in skin temperature. Skin warm receptors appear to exert no influence on sweating, since if internal temperature remains the same then widely different skin temperatures may accompany the same rate of sweating. Benzinger (1964) notes:

> Sensory warm impulses from the skin exert no measurable influence on sweating. In retrospect it appears that warm receptors could not possibly exert such an influence without upsetting the thermoregulatory system and defeating its purpose. If sweating depended on warm impulses from the skin, then the declining temperature of the skin with every onset of strenuous exercise would curtail or eliminate the response of sweating at the time when it was most needed.
>
> When the skin is the area of attack for a vigorous cooling mechanism it is not simultaneously a suitable location for the controlled temperature variable in a delicate system of regulation. Temperature regulation in a home would not be improved if the warm sensor of the thermostat were sprinkled with water and cooled during periods of overheating.

However, it would be wrong to exclude completely the skin from our consideration of the sensory side of the heat loss mechanism. Although the sweating response does not depend upon warm receptors serving as sensors at the skin, there is evidence that cold receptors located there can exert an inhibition on sweating rate. It is possible that under some circumstances such receptors could serve a very useful purpose and in discussing this a fundamental principle of feedback control is introduced.

Let us assume that a human subject is not exercising unduly but none the less is sweating at a high rate and then suddenly he finds himself in a cool environment. It would take time before the new environment could influence his central temperature and in the meantime he continues to produce sweat at a high rate. This could have the effect of reducing his temperature to below normal. It would be more efficient if the system anticipated the drop in temperature and therefore immediately reduced the sweat rate when coming into contact with the cold environment. Cold receptors at the skin that inhibit sweating can serve just this end.

Where there are delays in systems often the most efficient thing to do is to anticipate a change, rather than to wait for it to occur and then take corrective action. In the case of room heating it would be better if the radiator were switched off just before the set-point were reached. In the case of sweating a significant delay occurs between the central command to the sweat glands and the cooling effect of the sweat. A basic requirement of control systems is that information should be transmitted as rapidly as possible so that the corrective action is always appropriate to the error. If information is delayed, by the time it gets to its destination it may be quite out-of-date. This can lead to overshoot and oscillation of the output. In the present case the cooling effect of sweat will be unwanted if the body temperature has already reached normal. If temperature were to fall below normal, heating mechanisms would have to come into effect. A way of correcting oscillatory behaviour is to employ rapid information-transmission pathways that bypass the long delays.

Even if there were no skin cold receptors the situation would not be so bad as might be suggested here. In a cool climate it is not easy to loose heat in the form of evaporation and this means that the sweat would tend to remain on the skin, which is perhaps another good reason for inhibiting its production as soon as cold is encountered.

There is an aspect of temperature control to which control systems analysis is particularly relevant, and that is the subject of fever. It is believed (though not proven) that in fever, substances called pyrogens cause the hypothalamic set-point to rise. If this is indeed the case, then after the set-point has been increased an error would appear that would indicate that body temperature is too cool. The autonomic responses would then be put into effect in order to raise body temperature. While the output is approaching the new set-point, although the body temperature is in fact above normal, the subject shows all the signs of being cold, i.e. shivering and vasoconstriction. This all makes sense when looked at in terms of feedback control theory, but in the absence of such a framework it would be almost impossible to understand. In fact, there is another possible explanation, which is only speculative, and that is that in fever the gain of the feedback pathway is decreased. This could be brought about by a decreased sensitivity of temperature measuring neurons. Either way, an increase in actual temperature would be the response needed to restore equilibrium.

During hibernation a mammal seems to adopt a new set-point for the temperature controller at about 6 °C instead of 37 °C (Milsum, 1966). If the external temperature attempts to bring the body temperature down to a value below 6 °C, then heat production is increased to return the body temperature to the new set-point. By re-setting the set-point, heat production requirements are minimised, which leads to the question of why the set-point should normally be as high as 37 °C. Milsum proposes that having such a high set-point makes the animal better able to control its temperature in a hot environment. A lower set-point would place greater demands upon heat-dissipating mechanisms under such circumstances.

It is known that just as some species exhibit a seasonal variation in body temperature, so also is there a diurnal variation. At night, when we are sleeping, our body temperature is somewhat lower than it is during the day, even if we remain inactive. This raises the question as to whether or not we can account for the diurnal rhythm in terms of a change in the set-point. At present there is no satisfactory answer to this question, a matter that is discussed by Oatley (1974), and also in Chapter 8 in the context of thirst.

FEEDBACK CONTROL AND INTRINSIC FEEDBACK

Figure 42(a) shows a model of a fluid container with an outlet, it could be the stomach or a metal container with a hole in the bottom, and Figure 42(b) shows a crude representation of the thirst–body fluid regulation system. The former is simply a physiological component or a piece of metallic hardware. If it is the stomach then it forms part of the transmission pathway of the thirst system, but whatever is the physical realisation of Figure 42(a), it seems clearly not to be a control system. By contrast, (b) clearly is a negative-feedback controller. It has a set-point and performs goal-directed behaviour, i.e. drinking in order to regulate the internal environment, a condition necessary for the continuance of life. Yet despite the quite different nature of each system, they share at least one important characteristic: a feedback loop may be identified in each case.

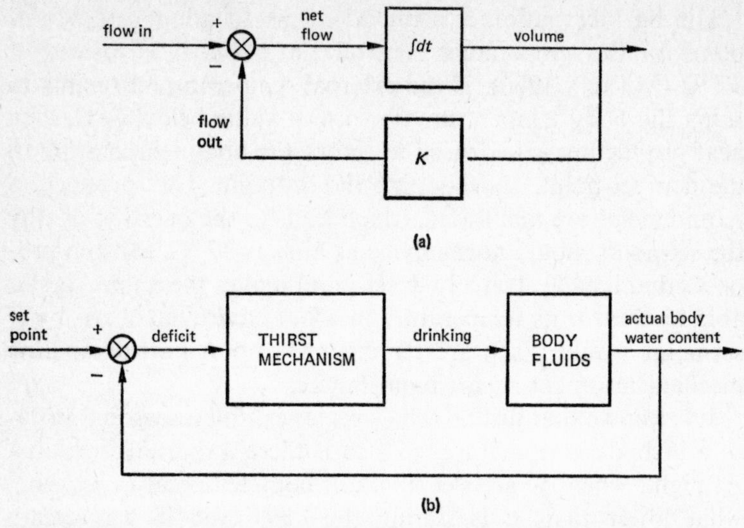

Figure 42 *Feedback control and 'intrinsic' feedback.* (*a*) *A fluid container such as the stomach or a bucket with a hole in the bottom exhibits 'intrinsic' feedback,* (*b*) *the thirst-body fluid system is a negative-feedback controller.*

This example serves to illustrate an important point, that at first may seem obvious but later presents fundamental philosophical problems. It appears that at times 'it just so happens' that feedback is present. A bucket of water with a hole in the bottom can only be mathematically characterised by including a feedback loop, as Figure 42(a) shows, and yet it would be stretching our imagination to the limit to call it a feedback control system. There seems to be no design or deliberacy about the feedback loop.

By contrast, a thermostatically actuated room heating system is clearly a feedback controller, in that an engineer has built it that way and has included a measuring transducer, a set-point, and error detector as part of the design. There is an intelligent being behind it and the feedback is intentional. Yet if we were to build mathematical models of two systems, one having 'intrinsic' feedback and the other being designed as a feedback controller, it might well be the case that the models would be identical.

The biological temperature control system appears to represent an example of deliberate feedback, like its engineering equivalent, but quite unlike the flow characteristics of the stomach. It attempts actively to match output to input. However, could we not say the same about the stomach? Could we not say that the object of the system is to match flow out to flow in? I think not, but I am unconvinced. After all, if we knew nothing about the stomach but merely measured the relationship between flow out and flow in, it would appear to establish a matching between these two. The problem seems to be in the realm of philosophy rather than in that of physiology or engineering, and I have no answer to it. Unlike engineering systems, we had no say in the design, and it is at this level that the problem is located. In a somewhat different context, Milsum (1966) has been troubled by the same problem, and reports:

> . . . in fact, it has been pertinently said, 'Feedback is in the eye of the beholder'. Thus, although population models, such as those of simple growth, warring armies, and insect ecologies, have been termed passive in the sense that one cannot confidently attribute a causal motive to the feedback channels without incurring difficult teleological problems, nonetheless mutually causal (feedback) paths can easily be identified.

POSITIVE FEEDBACK

The homeostatic systems of the body are built such that any error between the reference value and the actual value causes responses that result in error reductions, if not elimination. When body temperature falls below normal, heat is generated to return the temperature to normal. The sensation of thirst is provoked when there is a deficit in the body fluids, and if water is available then the animal corrects the deficit. We call such systems negative feedback because error tends to be self-eliminating as opposed to being self-generating.

Suppose, though, that in some mysterious way the temperature control system were to be 'wired up the wrong way', and instead of generating heat, a subnormal temperature formed the stimulus for the activation of heat loss mechanisms. If we were to give the unfortunate owner of such a system an ice-cube to eat, this would create an error in his temperature control system. The error would be the stimulus for heat loss, the heat loss would then exacerbate the error which would then increase heat loss, and so on until the subject dies from hypothermia. This would be an example of positive feedback, i.e. error serves to create an even larger error. Now obviously the body does not work in such a way, since the homeostatic mechanisms have evolved for the survival of the organism and are, of course, 'wired up' correctly. However, it can happen that positive-feedback performance not unlike that described can be observed in certain pathological conditions where homeostatic regulation may be over-ridden.

Figure 43 illustrates the heart's pumping effectiveness and, as we can see, the human heart normally pumps about 5 litres of

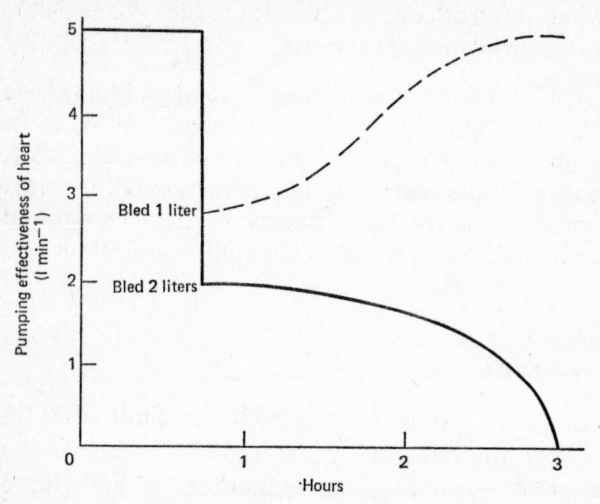

Figure 43 A case of positive feedback. When 2 *litres of blood are removed, death occurs in about* 3 *hours. Following loss of* 1 *litre the body is able to restore the circulatory efficiency to normal.* (*Source:* Guyton, 1971)

blood per minute. If the subject then loses 2 litres of blood, the amount of blood remaining in the body is insufficient for the heart to pump effectively. Arterial pressure falls, and this has the effect of reducing the flow of blood to the heart muscles via the coronary vessels. The heart is made weaker by this, with the result that pumping suffers still further which in turn weakens the heart. After three hours the pumping ability of the heart has reached zero. While all of this was happening the homeostatic feedback loops would not have been idle, but rather they would have been actively trying to restore normality. None the less, the positive-feedback effects dominated. It may be seen from the dotted curve in Figure 43 that when only 1 litre of blood is lost, the negative-feedback loops are able to dominate and pumping ability is restored to normal.

There is a feedback control system which attempts to restore the hearts pumping ability to normal. This is what we would accept as being a homeostatic controller and is of course a negative-feedback controller. However, in addition there is an 'intrinsic' positive-feedback loop.

We can in fact employ the discussion of positive and negative feedback to give a particularly good example of where systems of a totally different nature none the less have similar performance characteristics. In the one case the system is a feedback control system and in the other case an example of 'intrinsic' feedback.

Figure 44 shows a lever that is free to swing around a pivot. In (a) it is hanging vertically downwards, and we then introduce a disturbance to the system by displacing the lever to the position shown by the dotted line. The 'error' causes a force that

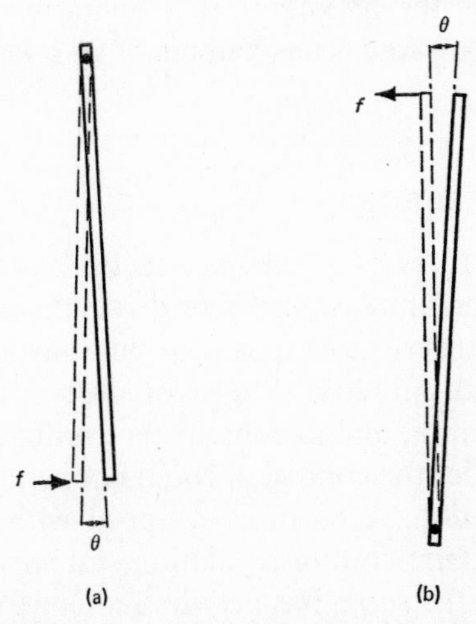

Figure 44 The difference between a stable and an unstable system. (a) Any disturbance that produces a displacement θ from the equilibrium position also produces a correcting force f. In other words this is a stable system. (b) This shows an unstable system. Any disturbance that produces a displacement θ is accompanied by a force f that increases the displacement from equilibrium.

attempts to return the lever to its original position. In other words it is a negative-feedback system – error produces corrective action that eliminates error. It is like swallowing an ice-cube and then observing that the drop in temperature puts heat producing mechanisms that return body temperature to normal into effect. In (b) the lever swings through a pivot at the bottom, but at the moment is precariously balanced in its equilibrium position. If we were to give it a displacement from its equilibrium position, then a force would be introduced that would take it still further away from its starting point. The larger displacement would then create a still greater force, and so on. This is a positive-feedback system, and is analogous to the temperature control system that is wired up the wrong way so that cooling is the stimulus for heat loss.

The lever is in a position of stable equilibrium in part (a), meaning that the lever tends to return to its equilibrium position following a disturbance. Part (b) shows unstable equilibrium, in this case the lever does not return to its equilibrium position. Stability is a concept that the control engineer is familiar with; a system such as the hypothetical temperature system that is wired up the wrong way is an example of an inherently unstable system.

STEADY-STATE ERRORS

The essential features of a negative-feedback system are an input, output and error detector. Error drives the output in such a way as to eliminate error. At first sight this may appear as somewhat of a logical paradox. Can error ever bring output into alignment with input, and in consequence reduce itself to zero?

Let us reconsider the control of room temperature in order to answer this question. A room is equipped with a thermostatically controlled, temperature-regulation system which has been switched off and the room has acquired a temperature of 10 °C as a result of the equilibrium established with the outside environment. A person then enters the room, switches on the temperature controller and sets the desired value to 20 °C. Will the room temperature ever reach 20 °C? The answer is dependent upon the characteristics of the complete system. We will at first consider the somewhat unrealistic case of a room with such perfect insulation that we can ignore all heat loss. The system is shown in Figure 45(a).

A comparison is made between the reference value and the actual temperature, the difference being the error. A positive error means that actual temperature is below the reference temperature and this forms the stimulus for heating the room. A negative error is the stimulus for cooling. In the present case there is an error of $20 - 10 = 10$, and heat is generated. The rate of heat production is proportional to the error, as shown in the diagram, i.e.

$$\text{rate of heat production} = K_1 \times \text{error}$$

If we integrate the rate of heat production we arrive at the heat quantity in the room. The room characteristics determine the relationship between the temperature of the room and the amount of heat present in the room.

The heat that is now being generated will increase the temperature of the room. This increase causes a corresponding reduction in the error and a reduction in the rate of heat production. However, as long as an error is present, no matter how small, heat will continue to be produced. This means that ultimately the room temperature is bound to reach 20 °C. When this occurs the error is zero and the rate of heat production is brought to zero. A graph of actual temperature and desired temperature is shown in Figure 45(b), where it may be seen that for a step input the output rises exponentially. The exponential response arises because output is proportional to error, and error is reduced as the output increases. In the steady-state, output and input come into alignment, and we would say that there is no steady-state error present. Up until this point an error is present which is known as a dynamic or transient error.

Let us now make the situation more realistic and suppose that heat is being lost from the room. We will assume that heat loss occurs at a rate proportional to the difference in temperature between the room and the environment, and we will also make the assumption that the environment temperature remains constant at 10 °C. Thus

$$\text{rate of heat loss} = K_2 \times (\text{temperature of room} - 10)$$

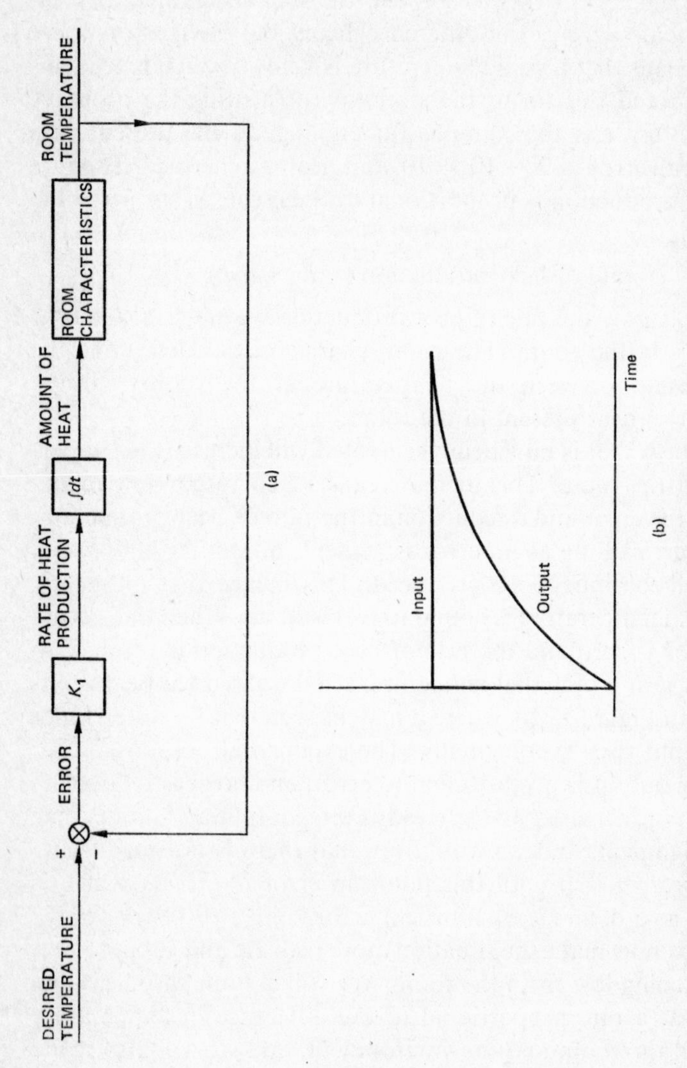

Figure 45 ***A room heating system where the rate of heat loss is zero. (a) The construction of the system, (b) the response of the system showing that output finally matches exactly the input.***

where K_2 is a constant. We can now modify Figure 45(a) to include heat loss, as shown in Figure 46(a). Rate of heat loss is proportional to the temperature gradient, and net rate of heat production is given by rate of heat production minus rate of heat loss.

When we start the system going with a reference value of 20 °C there will again be an initial error of 10, and then the temperature of the room starts to rise. It will continue to rise until the rate of heat loss is equal to the rate of heat production, at which point the temperature will have reached its maximum. Will this maximum be the reference value of 20 °C? No, it will be less than 20 °C. If we refer to Figure 46(a), we will see that for the temperature to be constant the amount of heat present must also be constant. This means that in the steady-state the net rate of heat production is zero. However, since the temperature of the room is higher than the environment, there will be a constant rate of heat loss and this necessitates a constant rate of heat production to compensate. As Figure 46(a) shows, rate of heat production is given by $K_1 \times$ error, so this means that an error must be present even in the steady-state.

If we were to plot a graph of input and output it would be as shown in Figure 46(b). There would be an exponential rise in the output but it would never attain the same value as the input. The disparity between the input and the output existing in the steady-state is known as the steady-state error. Note that the integrator remains but it now integrates a different quantity, i.e. production rate minus rate of loss rather than production rate.

The system characteristics determine whether or not a steady-state error exists. By introducing a rate of heat loss we have changed the system from what is known as an integral to a proportional controller. In the integral controller, output is proportional to the integral of error, whereas in the proportional controller output is proportional to error. More will be said on this subject in Chapter 6 when we discuss the mathematical analysis of feedback systems.

Cooling as well as heating may be applied to this particular room, so let us examine what would happen if the inhabitant for some strange reason decided that he wanted a temperature of

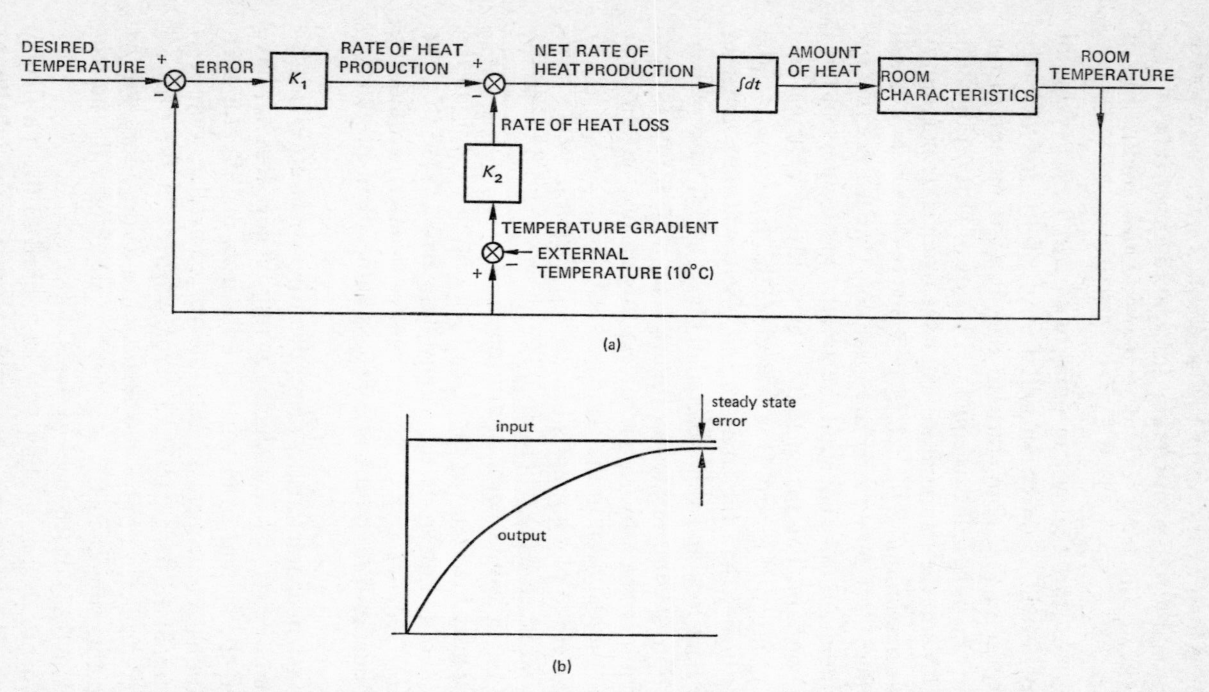

Figure 46 *A room heating system in which there is a rate of heat loss between the room and the environment. (a) The construction of the system, (b) the response showing the steady-state error.*

0 °C. Again we will assume that the environmental conditions have remained constant, and so we start out from a temperature of 10 °C. There is an initial error of 0 °C – 10 °C = –10 °C. The negative error is the stimulus for cooling the room, in other words there is a negative rate of heat production. However, as soon as room temperature falls below 10 °C heat is gained from the environment, in other words rate of heat loss takes a positive value. The result is that the temperature falls until the rate of heat gain from the environment is equal and opposite to the rate of heat removal. Since rate of heat removal is proportional to the error, there must remain a negative error in the steady-state, meaning that the room temperature will be above 0 °C. Figure 47 shows the input–output relationship of the system, and it

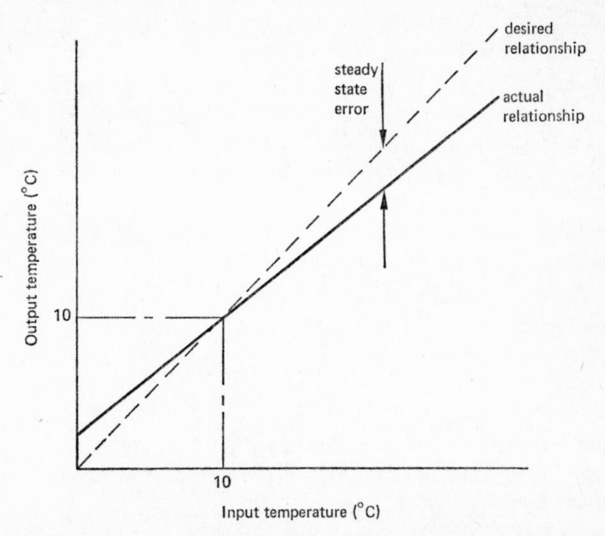

Figure 47 The steady-state performance characteristic of the system shown in Figure 46. Note that the steady-state error changes sign at 10 °C.

may be seen that the steady-state error changes sign at 10 °C, which is the point at which the effort changes from one of heating to one of cooling. No effort is required at 10 °C and the error is zero at this point.

The Accommodation Control System of the Human Eye In this section it will be argued that an analogy may be drawn between the temperature control system shown in Figure 46(a) and the accommodation control system of the human eye. The reader must be warned that it is only a theory of accommodation that is presented, the intuitive attractiveness of the model should not be confused with a concrete proof. The theory has been developed in detail elsewhere (Toates, 1970, 1972a).

The power of the eye's lens changes so that objects situated at various distances from the eye may each in turn be brought into focus at the retina, this activity being known as accommodation. In Figure 48(a) a distant target is in focus and so the lens is relatively flat. When, as in Figure 48(b), the object is brought

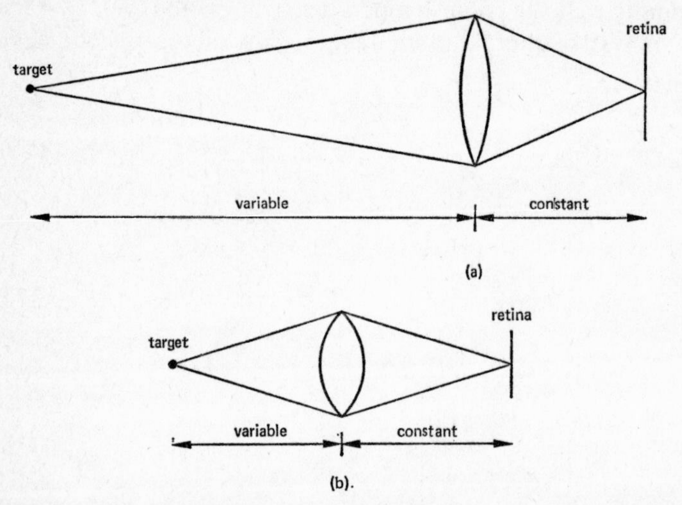

Figure 48 The accommodation of the lens. (a) A distant target is in focus and so the lens is relatively flat, (b) a near target is in focus and so the lens is relatively curved.

near, the lens must increase its power for the image to remain in focus. The power of the lens is automatically adjusted as the target distance is altered.

Accommodation is measured in dioptres. If the target is at $\frac{1}{2}$ metre then 2 dioptres of accommodation are needed in order to bring the image into focus at the retina.

$$\text{accommodation required (dioptres)} = 1/\text{target distance (metres)}$$

To focus at infinity therefore calls for 0 D of accommodation. The actual accommodation of the eye may or may not match that demanded on the basis of target distance, in practice the eye may be under or over-accommodated. We could consider the accommodation that is required in order to bring the target into focus to be the input to the control system. The actual accommodation of the lens forms the output. If these are different the eye is either under- or over-accommodated, and the defocus that results then serves to actuate the accommodation control system. Defocus corresponds to error.

Figure 49(a) shows a representation of the accommodation control system. Defocus is the error signal which gives rise to a

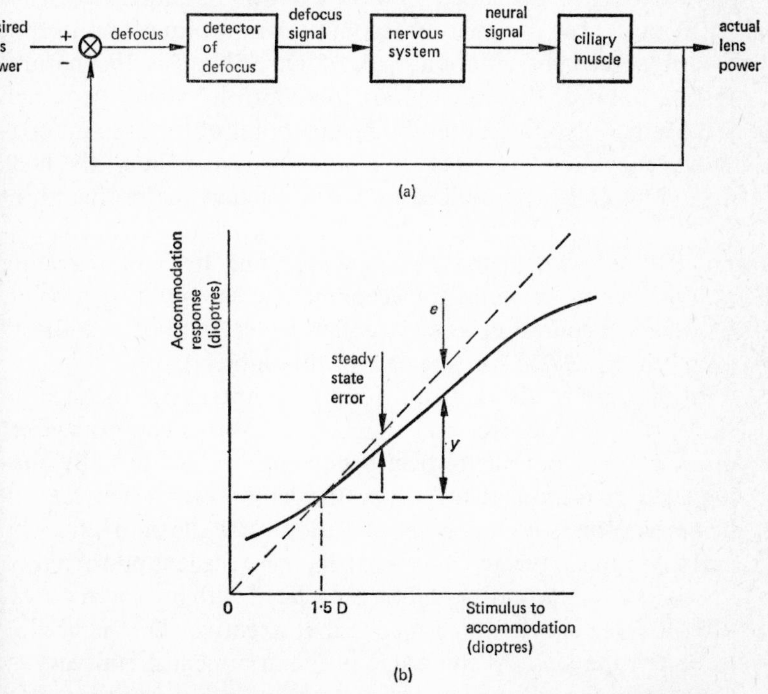

Figure 49. The accommodation control system. (a) A model showing the essential components of the system (source: Toates, 1972a). *(b) A typical stimulus–response curve for the steady-state performance of accommodation. The dotted line shows the relationship of perfect equality, while the continuous line is the actual relationship. The output y is measured from the resting point at* 1·5 D. *The steady-state error e is the difference between input and output. (Source: Toates,* 1972c).

neural innervation of the ciliary muscle. The ciliary muscle then determines the power of the lens.

In order to understand how the accommodation system works it is necessary to examine the input–output relationship that corresponds to the one shown for temperature control (Figure 47). It may be seen that Figure 49(b) has the same form as the temperature relationship shown in Figure 47. In Figure 47 input and output are in agreement at 10 °C, no steady-state error being present. This is because 10 °C is the temperature that the room arrives at in the absence of any active control. Therefore when the reference value is set to 10 °C it is not necessary to exert any effort, since this is the natural temperature of the room. Steady-state errors are associated with the exertion of effort. As Figure 49(b) shows, in the case of accommodation, input and output are in agreement at about 1·0–1·5 D. So is this the natural resting point of the accommodation control system? Although it is commonly taught that the resting point of the accommodation control system corresponds to a focus at infinity, the facts do not support this. Indeed they suggest that the resting point is at 1·0–1·5 D. One might expect that in darkness or in a bright but featureless field the system would find its natural resting point, since no stimulus for accommodation would be present. Under such conditions the eye comes to rest at about 1·0–1·5 D (see Toates, 1972a, for a review of this subject).

In the case of the control of room temperature, the steady-state error changes sign at 10 °C because at this point the effort must change from one of heat production to heat loss. By analogy, effort is needed to reduce the lens power when the accommodation is less than about 1 dioptre, while effort must be exerted in the opposite direction to increase the lens power above this value. We can call the effort associated with increased curvature positive effort, which means that negative effort is needed to flatten the lens. At this stage in the argument it is necessary to consider the autonomic nervous system in order to propose a neural basis for the dual innervation.

As its name implies, the autonomic nervous system controls the involuntary functions of the body. The system is composed of two divisions: the sympathetic and parasympathetic. Their action is antagonistic; for example, the sympathetic increases the vigour of the heart's pumping while the parasympathetic slows it down. The resultant effect within each organ is dependent upon the equilibrium established between the two divisions. Accommodation is controlled by the autonomic nervous system and, since it was believed that accommodation is at rest when focused at infinity, this required effort to be exerted in only one direction and consequently required only one autonomic division. It was known that the parasympathetic division was involved in increasing the power of the lens, and a sympathetic contribution was therefore ruled out. However, attitudes change, and corresponding to the recognition that the resting point is probably not at infinity, new results and interpretations emerged which demonstrated a sympathetic innervation of the ciliary muscle (see Cogan, 1937, for an excellent discussion). In view of our present understanding it seems reasonable to propose that increased sympathetic innervation produces a flattening of the lens to a power less than 1·0–1·5 D, whereas activation of the parasympathetic supply increases the power of the lens above this value. At the resting point an equilibrium is established between the tonic values of each division.

The input–output curve of accommodation shows that we are dealing with a proportional controller, because an error must exist to obtain an output other than one corresponding to the resting point. Since effort derives from the steady-state error, we would expect this also to change sign at about 1·0 D, which is indeed what happens. The eye is under-accommodated for near objects but over-accommodated for distant ones. It is concluded that the error signal accompanying over-accommodation is the stimulus for activating the sympathetic system, and an under-accommodated image produces an increased parasympathetic innervation.

A MODEL OF AN AUTONOMIC EFFECTOR CONTROL LOOP

The accommodation control system may be taken as a particular example to illustrate a theoretical model of an autonomic control loop having a more general application (Toates, 1972b). Such a model is presented in Figure 50(a). It is a simplification since only one division of the autonomic nervous system is

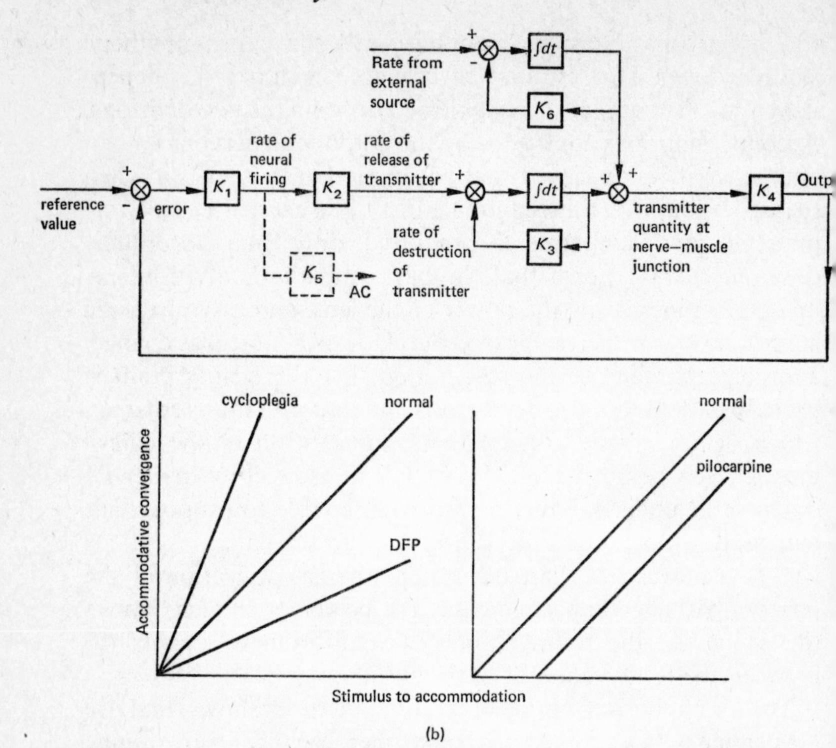

Figure 50 Systems analysis of an autonomic control loop. (a) Proposed model of the system, (b) the effect of various drugs on accommodative convergence. It may be seen that cyclopegic drugs and DFP change the accommodative convergence/accommodation ratio whereas pilocarpine leaves it unchanged. (*Source: Toates*, 1972b).

shown, but the general principles emerge from this model and they may be adapted to fit both divisions.

The difference between the reference value and the output gives rise to an error signal. The term K_1 relates the error signal to an innervation of the muscle. In the case of accommodation the defocus signal representing under-accommodation produces a parasympathetic innervation to the ciliary muscle. The terms such as K_1 should not be thought of as being linear operators. In all probability they are non-linear.

The transmission of information at the nerve–muscle junction is by chemical means. In the case of the parasympathetic nervous system, acetylcholine is the transmitter employed. In

Figure 50(a) gain K_2 relates rate of release of transmitter to the rate of neural firing. Rate of release of acetylcholine may be compared to the rate of heat production in the case of room temperature control. If we integrate rate of release we obtain total acetylcholine quantity at the nerve–muscle junction. Comparable to the rate of heat loss from the room, there is a destruction rate of acetylcholine which is made proportional to the amount of acetylcholine present, i.e. gain K_3. This then gives a net rate of production of acetylcholine, which forms the input to the integrator. Because of this feedback loop, in order for a fixed amount of acetylcholine to be present at the nerve–muscle junction there must be a constant rate of release of transmitter. This necessitates a constant rate of neural activity and hence a steady-state error.

The gain K_4 relates the output to the transmitter quantity, and represents the effect of the muscle. Acetylcholine, or a substance having similar properties, may reach the muscle-ending via the blood stream or by local application. Either way the total transmitter activity is the sum of that arising from neural activity and that introduced by other means. Transmitter originating from an external source will also decay at some characteristic rate, and this is accounted for by operator K_6.

Sympathetic control may be represented in a similar way to parasympathetic, and may crudely be compared to the cooling effort of the temperature control system. The sympathetic transmitter, adrenaline, is released into the blood stream as well as being released at the nerve-ending, and would therefore also have an effect via 'rate from external source'. In the case of accommodation it is known that the parasympathetic is the dominant division (see Toates, 1972a, for a discussion of this subject) and so where the following discussion refers to accommodation, only the parasympathetic division is considered. For the topics being considered this simplification probably does little injustice to the system, though in other respects we cannot consider the parasympathetic in isolation, even for accommodation.

The model may be employed to illuminate our understanding of certain drug effects. In the system shown in Figure 50(a), drugs may interact in at least two different modes: additive and multiplicative. If the drug adds to the effect of the transmitter released from the nerve-ending, it forms an input at the point marked rate from external source. The other possibility, that the drug acts multiplicatively, involves a change in the effectiveness of a given rate of release of transmitter coming from the nerve-ending. This implies a change in gain at either K_3 or K_4, depending upon the drug concerned.

As an index of the action of a drug, the effect that it has on accommodative-convergence can be observed. Apart from producing the accommodation response, accommodative-effort also produces convergence, an inward rotation of the eyes so that they continue to point at the target as it is brought nearer. In Figure 50(a) the operator K_5 relates accommodative-effort to accommodative-convergence (AC). Certain drugs have the effect of increasing the forward loop gain. If K_3 is decreased, i.e. the rate of destruction of transmitter is slowed down, then a smaller rate of release is needed in order to obtain the same transmitter quantity at the nerve–muscle junction. This means that the neural innervation is reduced and a corresponding reduction in accommodative-convergence is observed. In other words, decreasing K_3 should lower the accommodative-convergence to accommodation (AC/A) ratio. Drugs that retard the destruction of acetylcholine should have the effect described, diiosopropyl-flurophate (DFP) being an example of such a drug. Figure 50(b) shows the drug's effect. The AC/A ratio is linear as before application, but it takes a lower value, as the model would predict. The diagram also shows a cycloplegic drug, which hinders the effect of acetylcholine on the muscle, having the opposite effect. In terms of the model, a cycloplegic drug lowers gain K_4. The effect of both drugs is multiplicative in that it is necessary for acetylcholine to be present for the drug to have any effect. The effect is proportional to the amount of acetylcholine present.

An example of a drug that would be expected to have an additive as opposed to a multiplicative effect is pilocarpine, since it mimics acetylcholine and therefore acts as an independent stimulus. In terms of the model, it forms an input at the point marked 'rate from external source'. Pilocarpine would attempt to increase accommodation by a fixed amount, but this would be opposed by an equal and opposite decrease in accommodative-effort. Since accommodative-convergence is proportional to accommodative-effort the appropriate equation is

$$AC_n = AC - K$$

where AC_n = accommodative-convergence after the effect of pilocarpine, AC = accommodative-convergence before pilocarpine, and K = a constant.

The corresponding equation for the effect of DFP is

$$AC_n = AC \times K$$

As shown in Figure 50(b), pilocarpine has an effect independent of the level of accommodation.

If the model is extended to other systems, certain drug effects that might otherwise have remained obscure may then be understood. Working with the pupil, Swan and Gehrsitz (1951) found that if pilocarpine was administered prior to DFP the reaction to DFP was almost abolished. Normally DFP produces pupillary constriction, as does pilocarpine. Pilocarpine would be expected to produce a counteracting reduction, if not total cessation, of the parasympathetic neural innervation to the pupil. There would be a corresponding reduction in the rate of acetylcholine production. As has been explained, the effect of DFP is dependent upon the presence of acetylcholine so it is to be expected that pilocarpine will reduce or abolish the effect of DFP. If the drugs are administered at the same time, pilocarpine appears to exert less inhibitory effect on the action of DFP. It is suggested that this is attributable to the fact that at the time of injection and, due to various lags, for a while after, there is still acetylcholine present whose decay may be retarded by DFP. Lowenstein and Loewenfeld (1953) claimed that the effects of physostigmine (a drug supposed to have the same properties as DFP) on miosis are additive rather than competitive. This conclusion is not really supported even by the authors' own results; in some subjects the combination is no more potent than physostigmine alone.

The model may also be used to explain certain features of an experiment of Guyton and Gillespie (1951) on arterial pressure regulation. Arterial pressure is controlled largely by the sympathetic nervous system, control being exerted by the constriction and dilation of blood vessels. In the experiment, a constant infusion of the sympathetic transmitter substance epinephrine was given both to normal dogs and to dogs with total spinal anesthesia such that all sympathetic neural transmission was blocked. A much larger rise in arterial pressure was observed in the subjects with spinal anesthesia than in the normal subjects.

For subjects with a functioning sympathetic nervous system the tendency of epinephrine to cause a rise in arterial pressure is detected by baroreceptors and these convey a signal to inhibit the sympathetic impulses sent to the blood vessels. In the situation caused by anaesthesia, such a reduction to counteract the increase in arterial pressure cannot occur. Spinal anaesthesia causes the loss of about 50 mmHg in arterial pressure that would otherwise be available to counteract the effect of epinephrine. The maximum arterial pressure was little different between the two situations; the greater effect during use of anesthesia is because arterial pressure starts out from a lower value.

OPEN- AND CLOSED-LOOP TESTS

In the example just given there is an information transmission pathway from the baroreceptors (that detect a rise in arterial pressure) to the source of the sympathetic impulses sent to the blood vessels. Thus when arterial pressure rises, a compensatory decrease in sympathetic stimulation occurs and this has the effect of bringing arterial pressure back to normal. In other words we are dealing with a closed-loop system. However when anesthesia makes a break in the loop we convert a closed-loop system into an open-loop system.

In the normal animal, when we inject epinephrine and observe the effect on arterial pressure we cannot ignore the feedback effect via the baroreceptors. But when we have made a break in the information transmission pathway, the effect of epinephrine can be studied in isolation and no corrective action can be taken. If we were interested in the relationship between epinephrine injection rate and arterial pressure, then obviously the latter would be a simpler situation to analyse, though of course not so easy to produce. For the sake of the relative ease of analysis one often goes to considerable lengths to find how the loop may be broken. Normally both open- and closed-loop tests are carried out and the information derived from each is compared.

The problem of observing the response to a stimulus without allowing the response to influence the system under examination is by no means new. The famous French physiologist Claude Bernard, and several others, have employed what we would now call open-loop tests to study drinking. The problem in the case of thirst is as follows.

When an animal is thirsty and it is allowed water to drink, the water changes the state of the body fluids and hence alters the animal's thirst. The reader will recognise that we are dealing with a closed-loop system. To study the animal's behaviour following, say, 24 h of water deprivation, then ideally we want to allow the animal to drink but to prevent drinking from influencing the animal's thirst. We might then be able to get some kind of steady-state measure of drinking behaviour at each of several levels of thirst.

Bernard, in his experiments, quite literally broke the loop. He cut the oesophagus of a horse and brought the upper end to the outside, so that any water that the animal drank poured out of the open end and failed to reach the body fluids. In such a situation it is found that an animal considerably overdrinks on the basis of what would be consumed normally, but some kind of transient satiety appears to be derived from the act of drinking itself. In other words not all sources of feedback are removed by this operation, though the most important is.

Fortunately, less drastic means are now available for opening the loop of the thirst control system. If a rat is deprived of water it will lick a jet of air just as it would lick the opening of a water spout to obtain water, the important difference being that the body fluids derive no benefit from air licking. Oatley and Dickinson (1970) found that the rate at which a rat would lick at a jet of air depended upon the level of water deprivation, the rate being sustained over a relatively long period of time. It was claimed that by this open-loop technique it is possible to obtain a measure of thirst in that air licking is an index of deprivation level.

In order to show where open- and closed-loop tests have proved to be of value in mathematically characterising a biological system we can again turn the discussion to the visual system.

Vergence Eye Movements Figure 51 shows the eyes viewing a target at position A and it may be seen that the image falls on

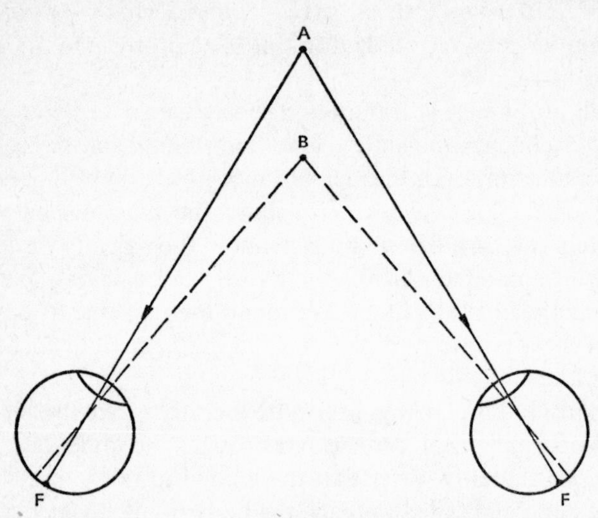

Figure 51 The need for a vergence function. The eyes are viewing a target at A *and the image falls on corresponding points in each eye. For ease of understanding they are called* F *to designate foveal position. The target is then moved to* B, *and as shown by the dotted lines the image no longer falls on corresponding points. Convergence is demanded.*

corresponding points (the foveas) in the two eyes. (The fovea is a small region of the retina having a very high resolving power. It is when the image falls here that visual acuity is at a maximum.) The dotted lines show that when the target is moved to B the image now no longer stimulates corresponding points. It will be to the right of the fovea in the right eye but to the left of the fovea in the left eye. The only way that alignment between image and fovea can be restored is for the eyes to turn inwards, i.e. the left eye to turn clockwise and the right eye anti-clockwise. Such a response would automatically result if the target were to be moved from A to B, and is known as a vergence eye movement. The stimulus for vergence in the present case is retinal disparity, in other words the image no longer falls on corresponding retinal points. We are again dealing with a negative-feedback system since disparity produces a vergence response that served to eliminate disparity.

Rashbass and Westheimer (1961) studied the vergence system and developed a technique that effectively opened the feedback loop. They imposed a disparity at the retina and measured the vergence response that the eyes made. However, instead of allowing the vergence response to eliminate the disparity, as would normally occur, the experiment was designed so that the disparity was held constant whatever the response. This was done by detecting the response and adjusting the disparity accordingly so that disparity was fixed. Fixed disparities of various magnitudes were applied at the retina and in each case the response was noted.

The experiment is somewhat analogous to the familiar story of a man on a cart holding a carrot out in front of a donkey. The donkey would normally be part of a closed-loop system, which would serve to eliminate the error in distance between the carrot and the donkey's mouth. Consequently, the donkey attempts to eliminate the disparity but of course in this particular case never succeeds, and the disparity remains constant. It might even be possible to quantify the relationship between disparity and velocity, but perhaps the analogy has been taken far enough already!

To return to vergence, Rashbass and Westheimer found that a fixed disparity produced a constant-velocity vergence movement, the relationship being linear over a wide range with a gain of 7–10 degrees per second per degree of disparity. On the basis of this performance characteristic we can determine the possibility of drawing an analogy between this system and the models of room temperature control that we discussed earlier.

Let us suppose that we disconnect the thermostat from the temperature control system, and in so doing open the loop. We then generate heat into the room at various constant rates and observe the effect on room temperature. For the system shown in Figure 45(a) the room temperature will steadily rise in response to a constant rate of heat production, since there is no heat loss from the room. However, in the case of the system shown in Figure 46(a) the temperature will rise only until the rate of heat loss from the room is equal and opposite to the rate of heat gain, at which point a stable temperature will be observed. We can therefore conclude that, in contrast to the accommodation system discussed earlier, vergence is analogous to Figure 45(a) rather than to Figure 46(a). In the case of accommodation it was argued that the system's characteristic arises from the fact that acetylcholine is being destroyed at the same rate as it is being formed. The different characteristic in the case of vergence should not be interpreted to mean that the transmitter is conserved, since in all probability the difference lies in the neural circuitry of the two systems, which accounts for the different relationship between error and output. Does vergence exhibit any other performance characteristic associated with controllers of the kind shown in Figure 45(a)?

With respect to the normal closed-loop situation, we have noted that in the steady-state there is no error between input and output for the system shown in Figure 45(a). Similarly, Westheimer and Rashbass report that the vergence system exhibits no steady-state error for a constant input. For a second comparison we can consider what would happen if we were to switch off the controller shown in Figure 45(a) as opposed to changing the set-point to a lower value. We would find that the temperature would be maintained exactly at the reference temperature even after we had switched the controller off. There is some evidence that vergence does this also – after removal of the target the eyes tend to retain their position of vergence for a long time (Alpern, 1969). To give another analogy, the vergence output appears to be like the charge on a capacitor. When the voltage source is disconnected the capacitor retains its charge.

The difference in the control system characteristics between accommodation and vergence reveals itself in ways that have long been familiar to the ophthalmologist, though control theory explanations are only just beginning to appear in the ophthalmological literature. It is known that vergence is more accurately controlled than accommodation, and Fincham and Walton (1957) reported that:

> . . . there is little tolerance to errors in convergence. We are very sensitive to diplopia and once fusion has been obtained by suitable

adjustment of convergence it is held on to firmly, as though the act of fusion locked the convergence mechanism with respect to the position of the retinal images. Accommodation behaves quite differently. Here there is a great tolerance to discrepancies . . .

Morgan (1968) also contrasted accommodation and vergence as far as accuracy is concerned.

For a review of the vergence control system and consideration of some of the problems and contradictions involved in applying systems techniques, the reader should consult the paper of Toates (1974a).

CONTINUOUS AND SAMPLED DATA SYSTEMS

The accommodation control system of the eye can serve as the starting point for the present discussion. Assume that accommodation is at rest, and then suddenly the target that forms the object of attention is brought nearer to the eye. After one reaction time (the time from the presentation of a stimulus to the appearance of a response) the accommodation of the lens will start to increase. If the target is now returned to its original position, one reaction time after the return movement of the stimulus the accommodation response will also start to return to its original position (Campbell and Westheimer, 1960). If a pulse stimulus of a length shorter than the reaction time is applied to accommodation (the target is brought nearer and then quickly returned to its original position), the stimulus will have returned to its starting point before the accommodation response even begins. None the less, the lens responds one reaction time after the beginning of the pulse but then changes direction after an interval of time corresponding to the pulse length. Campbell and Westheimer (1960) noted that accommodation appears to be a continuous monitoring system.

The important characteristic of the system is that information is transmitted continuously, and after the reaction time has elapsed the response that occurs is always appropriate to the stimulus applied one reaction time earlier. The same is true of the vergence controller (Zuber and Stark, 1968).

The saccadic eye movement system, to which we will now turn our attention, exhibits quite different performance characteristics from accommodation or vergence. The purpose of saccadic eye movements is to regain foveal fixation following either a sudden movement of the target or a voluntary attention shift to a new target located at the same distance from the subject. Thus it should, in terms of function served, be distinguished from vergence, which is applicable when the target is moved from near to far. In the case of the saccadic system both eyes rotate in the same direction, whereas for vergence they rotate in opposite directions. The saccadic eye movement system corrects for positional error, there is yet another controller that is responsible for velocity errors.

If the target is displaced suddenly so that a saccadic eye movement is called for, such a movement will be executed after a length of time corresponding to the reaction time of the system. However, if the target is then quickly returned to its original position a period of time much longer than the reaction time may elapse before the eyes make a saccade that returns them to their original position (Zuber and Stark, 1968). Information is not continuously transmitted but rather the controller appears to have a refractory period following a response during which the system is 'switched off'. Young and Stark (1963) presented the following evidence for the saccadic eye movement controller being a sampled data system:

1. The eye movement response to a target movement pulse of less than 0·2 s duration is a pair of equal and opposite saccadic jumps separated by a refractory period of at least 0·2 s. The response predicted by a continuous system with a pure delay would be a pulse of the same duration as that of the target, delayed in time by an amount equal to the reaction time.
2. Under open-loop conditions, in which the effective visual feedback is modified by addition of an external path from measured eye position to target position, the response to a target step is a staircase of equal amplitude saccades spaced approximately 0·2 s apart.
3. During constant target acceleration the eye velocity changes in rather discrete jumps at 0·2 s intervals, indicating the discrete nature of the pursuit system; position errors are corrected by saccades in the direction of target motion.
4. The frequency response of the system shows a peak in gain in the region of 2·5 cps(Hz), consistent with a sampled data system operating at a sampling period of 0·2 s.

5. The inaccuracy of saccades associated with anticipation of square-wave target motion indicates that such saccadic movements cannot be modified by any visual information appearing in the interval 0·2 s previous to the saccade.

Sampled data controllers, in which the error is not examined continuously but rather only looked at periodically, are familiar to the engineer. The output between samples is based upon the error occurring when the most recent sample was taken. A switch is closed at regular intervals for the error to be examined. A 'hold' component serves to maintain the output in the absence of an error signal. The engineer would employ such a sampler when more than one system needs to share the same communication channel and time has to be allocated to each system. It may be that the saccadic controller exhibits sampling because the neural pathways are being employed for some other purpose.

Probably the most interesting and at the same time least clear aspect of the saccadic eye movement system is precisely what determines when the switch will be closed. Young and Stark (1963) suggest that samples are taken regularly at every 0·2 s; this makes the system analogous to the kind of thing an engineer would build. However, Milhorn (1966), while acknowledging the sampled data mode of operation of the system, objects to the assumption that the sample is taken at fixed intervals. Milhorn claims that the restriction on saccadic movements is such that there is a minimum time between the responses, apart from this the time between responses may be variable.

It is perhaps not unreasonable to suggest that when an error above threshold appears, a response is executed and then the transmission of information is switched off for 0·2 s. If at the end of this time an error is still present then a second response occurs. It is possible that the system is constantly monitoring error but the sampling is nearer the response end of the system.

Control Theory and Behaviour

So far we have considered such topics as temperature control, arterial pressure regulation, pupil-size, accommodation and convergence. By now the reader whose interest is towards psychology rather than physiology may be wondering whether the nearest that we get to his or her area of interest is in discussing quasi-behavioural reflexes. I hope in this section to show that this is not the case, on the contrary that control theory is central to both the organism's perception and its locomotion in the environment.

Norbert Wiener (1948) is generally credited with being the first to recognise the parallel between engineering control systems and the activity of the nervous system. It is worth quoting Wiener's original account at some length, so that the subject may appear in its historical context:

Now suppose that I pick up a lead pencil. To do this, I have to move certain muscles. However, for all of us but a few expert anatomists, we do not know what muscles these are; and even among the anatomists, there are few, if any, who can perform the act by conscious willing in succession of the contraction of each muscle concerned. On the contrary, what we will is *to pick the pencil up*. Once we have determined on this, our motion proceeds in such a way that we may say roughly that the amount by which the pencil is not yet picked up is decreased at each stage. This part of the action is not in full consciousness.

To perform an action in such a manner, there must be a report to the nervous system, conscious or unconscious, of the amount by which we have failed to pick up the pencil at each instant. If we have our eye on the pencil, this report may be visual, at least in part, but it is more generally kinesthetic, or, to use a term now in vogue, proprioceptive. If the proprioceptive sensations are wanting and we do not replace them by a visual or other substitute, we are unable to perform the act of picking up the pencil, and find ourselves in a state of what is known as *ataxia*. An ataxia of this type is familiar in the form of syphilis of the central nervous system known as *tabes dorsalis*, where the kinesthetic sense conveyed by the spinal nerves is more or less destroyed.

Wiener continued:

The central nervous system no longer appears as a self-contained organ, receiving inputs from the senses and discharging into the muscles. On the contrary, some of its most characteristic activities are explicable only as circular processes, emerging from the nervous system into the muscles, and re-entering the nervous system through the sense organs, whether they be proprioceptors or organs of the special senses. This seemed to us to mark a new step in the study of that part of neurophysiology which concerns not solely the elementary processes of nerves and synapses but the performance of the nervous system as an integrated whole.

It is not difficult to recognise the negative-feedback control system components in Wiener's description. We have an input to the system that takes the form of some kind of neural representation of a particular place that the arm should be reaching to. The output of the system is the actual position that the arm occupies. Information on the arm's actual position is fed back via the retina and also via neural pathways from the muscles. The nature of the response is determined by a comparison between where the arm should be and where it in fact is. Is this realistic and is it possible to imagine a neurological embodiment of the control system components?

It is known that the cerebellum receives impulses from both the cerebral cortex and from the muscles (Ruch, 1951). The cerebral cortex is where voluntary movements originate, and so the input to the system must arise in the neurons located there. In other words the cerebral cortex has some kind of internal representation or model of the environment. The muscles are the source of feedback information that is transmitted to the cerebellum and so we might reasonably consider the cerebellum to be the site of the error detection operation. The command to the muscles is determined on the basis of a comparison between neural activity representing the desired position and other neural activity that represents the actual position of the limb.

When Wiener states that there are '. . . few, if any, who can perform the act by conscious willing in succession of the contraction of each muscle concerned', in principle he is describing the same problem that we discussed in connection with heating a room. I hope that I do not violate psychology terminology too much if open-loop room heating is called an S → R system and feedback control heating an S ⇄ R system, where S means stimulus and R means response. We noted that just so long as all of the characteristics of the system remained constant the S → R heating worked perfectly. To compare this system with muscular control, let us suppose that it were possible consciously to will a certain muscle to execute a specific response. Somewhere in the cortex a command signal would arise, this would set up activity in a motor neuron and the muscle would contract. This is an S → R system in that there is only one direction of information transmission. Such a system might almost work, provided that all the characteristics of its components remained constant, but if for example the muscle fibres were fatigued then a different response would be obtained for the same command. The system would of course be quite unable to do anything about this unless the command centre were informed of the muscle's weak response.

When the reader stops to think about the problem, he or she will probably find it impossible to imagine how the system could perform without some kind of feedback control either proprioceptive or by means of the visual system. It seems almost intuitive that our muscular activities are guided by a comparison between what we want and what we have. It is not that a particular set of muscle fibres is activated by a particular conscious stimulus, but rather effort is made to be dependent upon need as defined by the error. Thus if we have regularly picked up a certain object after having learned to estimate its weight, and then suddenly the object is made heavier, our nervous system does not admit defeat and pleads that the correct command was given. Rather the system recruits extra muscle fibres and increases the activity of those already in use. This is an automatic response to the new situation; consciously we only willed the act of picking up the weight and not the number of muscle fibres to be employed. To draw a comparison, if the weather changes, the S ⇄ R heating feedback system automatically changes the heat production rate to be appropriate to the new circumstances.

A good example of the negative-feedback control of posture may often be seen during the course of watching a 'Western' at the cinema. A horse without a rider is standing still, the muscle activity in the animal's legs being appropriate for its own weight. Suddenly the hero of the 'Western' has to make a quick getaway, and jumps quite without warning onto the back of his horse. The horse does not collapse as a result of the load and then only recover after sufficient time has elapsed to 'consciously' re-estimate the situation, which might be the case if the system were not so cleverly designed. Rather, as one can observe, the horse is transiently just slightly lowered by the additional weight. The slight change in the muscles is the stimulus for the innervation to the muscles to be increased so as to be appropriate for the new weight.

Ruch (1951) draws attention to the fact that feedback by neural transmission pathways introduces a relatively long delay into the system. The reader will recall, from the discussion of temperature control, that where delays are present in the system the problem of over-shoot and oscillation is liable to be encountered. Ruch suggests that stability is obtained in voluntary movements because the system exhibits a slow speed of response that compensates for the lag in the feedback pathway. As evidence, Ruch mentions the slow development of cortical discharge, which means that the system tolerates a transient error that results from the slow speed of response as being the inevitable price to be paid for stability. More will be said on the subject of stability later.

It would be quite wrong to apply an S → R model to the nervous activity of locomotion. It is not like flicking a switch and making the light come on, but rather it is like switching on the intention to make the light come on. Once this has been done the system then proceeds to achieve the desired end-product with a certain amount of flexibility, and is able to compensate for uncertainties. Stimulus–response accounts of behaviour seem to suggest that a particular environmental stimulus is associated with a particular afferent stimulation which in turn is associated with an efferent discharge and a muscular response, there being a fixed circuitry that is cemented by learning. This view is no longer acceptable.

For some nervous system functions, the S → R model is appropriate. There are innate reflex circuits, such as the one governing that when I touch a hot stove I move my arm away from the painful stimulus in a predetermined way. In this case a sensory pain receptor has a fixed connection with a motor neuron, which makes the system somewhat like the lamp and switch circuit. However, with the exception of examples of this kind, the locomotion of the human in his environment is to be understood in terms of guidance and modification of responses according to the consequences of the response.

It is probably not too much of an over-simplification to suggest that there has been a tendency for organisms to evolve from open-loop ($S \rightarrow R$) to closed-loop ($S \rightleftarrows R$) control. Thus the behaviour of a frog consists largely of stereotyped responses to environmental stimuli. Its visual system is constructed to 'see' and respond to such aspects of its environment as moving flies, and it may well starve to death surrounded by dead flies. Similarly the pecking behaviour of chicks is released by small objects in the environment, and if the visual world of the chick is distorted it is quite unable to adapt to the distortion. By contrast, man is able to learn to adapt even to an inversion of his visual world produced by prisms. The nervous system of man has plasticity in that a number of possible strategies may be employed in matching actual performance to desired performance.

The principles of feedback may be observed in a wealth of nervous system activities, some perhaps not so obvious as the movement of limbs. Speech is an example of feedback control, though one might at first have imagined that it was brought about by a one-way flow of information from the speech areas of the cortex to the mechanical parts of the vocal apparatus. If one imposes a pure delay in the feedback pathway, by holding a microphone at the subject's mouth, recording his speech on magnetic tape and then playing it back through earphones but delayed by 0·2 s, subjects will find it almost impossible to talk (Fitts and Posner, 1967). If they are able to say anything at all then it will be in the form of a stutter.

If, on the other hand, instead of delaying information the experimenter employs intense noise in order to prevent subjects from hearing their own voice, then only a mildly disturbing effect on speech can be observed (Fitts and Posner, 1967). Though these procedures disrupt external feedback they cannot eliminate vibrations through the head bones. The nervous system in general is very good at extracting information from the most subtle of cues, but to impose a delay in the feedback pathway presents quite a different problem. The reason that the task is so difficult is that the subject is normally receiving reports informing him of what he has just said, and these reports allow him to proceed. A comparison can be made between what he wants to say and what in fact he has said. When the report is late it is as though he has not said the word, and so he tries again. An unintelligible stutter results, the result being recognisable to the engineer as a form of instability.

One might expect a major source of confusion to the subject to be the incompatibility between the delayed external feedback and the undisturbed internal feedback through the head bones. In the case of loud noise this confusion is not present.

This is probably the most appropriate place to discuss in detail the application of control theory to specific problems, and then to see if any general principles emerge from the cases considered.

TRACKING

In order to keep a car at a reasonable distance from the edge of the road, the driver makes observations of the car's actual position. He then proceeds to compare the car's actual position with his notion of where the car should be. The corrective action that is executed via the steering wheel is dependent upon the outcome of this comparison. In other words, the driver is the error detector of the system.

When we come to examine the system more closely we can recognise some of the more specialised characteristics of control systems that are familar to the engineer. For example, the driver does not at all times devote his visual attention exclusively to observing the position of the car. He will at intervals divert attention to his rear-view mirror and to the various dials on the panel. This means that the position of the car, a continuous function, is determined by information extracted at intervals – in other words we have an example of a sampled data controller. The mode of the controller will however depend upon the quantity of information that needs to be handled. When the road is straight the driver may be able to afford the time to select a radio station, but where there are bends in the road his visual system will be devoted to estimating the position and speed of the car. The ability to divide attention, that is required when driving a car, is impaired by cannabis (Casswell and Marks, 1973) as well as alcohol. As will be seen in a moment, even when concentrating on the car's position the driver is still behaving as a sampled data system, so normal car driving is to be seen in terms of relatively voluntary and somewhat random samples (for example, the time spent looking in the mirror) superimposed upon a basic sampled data mode of control.

When a learner motorist starts to drive, the movement of the vehicle is somewhat erratic. The driver finds himself coming too near to the kerb and therefore swings the car outwards. This corrective response tends however to overcompensate and so he then forces the car in to the kerb again. Compare this performance with the smooth responses executed by the skilled motorist, and it seems that we are dealing with a pure error-actuated controller in the case of the novice and a more sophisticated controller involving learning and computation of velocity information in the case of the skilled motorist.

A formal theory to account for tracking behaviour was developed by Craik (1947) and was accompanied by controlled experiments in which the only demand upon the operator was to track. Sadly this work did not reach fruition before Craik's death, but a theory had already emerged, and the following is an attempt to present its essentials.

Craik concluded that the human operator behaves as an intermittent correction controller. This conclusion was based upon the fact that a time record of tracking errors shows a wavy form with a principle period of about 0·5 s. Such periodicity cannot be attributed to the time needed for the error to reach threshold, since changing the gain of the system by amplifying the error has little or no effect. The sampled data mode of control has been confirmed by others (see, for example: Lange, 1967, and Welford 1968).

The corrections are essentially ballistic. In other words, once the correction has been computed from error information the response is executed in a predetermined (open-loop) manner. If an error is presented to the operator and then his eyes are screened just before the corrective movement is made, his response will still be accurate to within about 10 per cent. Craik argued that activities such as playing musical instruments are carried out at such a speed that they would be impossible to perform if they were continuously dependent on feedback with the inevitably long reaction times.

Craik was able to summarise man as a control system component involving the following four items.

1. Sensory devices, which transform a misalignment between sight and target into suitable physiological counterparts, such as patterns of nerve impulses, just as a radar receiver transforms misalignment into an error voltage.
2. A computing system which responds to the misalignment-input by giving a neural response calculated, on the basis of previous experience, to be appropriate to reduce the misalignment; this process seems to occur in the cortex of the brain.
3. An amplifying system – the motor-nerve endings and the muscles – in which a minute amount of energy (the impulses in the motor nerves) controls the liberation of much greater amounts of energy in the muscles, which thus performs mechanical work.
4. Mechanical linkages (the pivot and lever systems of the limbs) whereby the muscular work produces externally observable effects, such as laying a gun.

According to Craik's account, correction is based upon error but not guided by it. Skill involves learning responses to be applied to each situation. If the system were to be guided at all times by error and the response to be modified during its course on the basis of error, then this would demand a very slow speed of response. The delay in the transmission pathway is an inevitable source of instability, in that by the time the response has been made it is out-of-date. In these terms we may understand the difference between the novice and the skilled motorist. The beginner has no stored programme of action to guide his responses, but rather he responds on the basis of error alone. The delay in the system means that for some time after the error has been eliminated the driver is still taking the same corrective action, which can of course result only in an error of the opposite sign being produced. The response in terms of the excursion of the steering wheel is not the optimum for the particular error present. By means of trial and error (note that this expression has a control theory flavour about it) the skilled motorist has been able to learn the required optimum characteristics to employ. In addition, he utilises velocity information. This means that his response is not only the optimum on the basis of error but also on the basis of what error is likely to be in the future. Thus, when the motorist approaches a bend he steers into it, thereby showing anticipation of the curve to come.

Although the beginner is probably not so naive that he is employing only pure error information and no anticipation, none the less, improvement in performance with time can be described as a gradual shift from predominantly error control to error control plus anticipation in an open-loop fashion.

A similar phenomenon may be observed under quite different circumstances in the accommodation and vergence control systems that have already been introduced. These are error actuated; in the case of accommodation, defocus corresponds to error, while for vergence, error takes the form of retinal disparity. If suddenly the target distance is changed, an error appears and after a period of time (the reaction time of each respective system) the accommodation and vergence outputs will be driven to new values thereby eliminating the error. However, if instead of employing step changes, the target distance is made to vary in a smooth and predictable way (for example, sinusoidally), then the response appears ahead of where it would be calculated to occur on the basis of the step response. Consequently, the error is very much reduced (Rashbass and Westheimer, 1961; Stark, Takahashi and Zames, 1965).

To return now to the subject of tracking, it is probably worth mentioning that the stimulus for much of the research on this subject has come from engineering. The engineer's interest is to optimise a system that man forms a part of. In practice, tracking tasks generally fall into one of two classes: compensatory and pursuit tracking.

As hard as I have tried I cannot find a particularly good example of a pure compensatory tracking task that is of relevance in peacetime. Consequently, since a film such as 'The Battle of Britain' is so rich in examples, it is to here that we turn. Typically, the pilot of a Spitfire spots a Messerschmitt and attempts to fly his plane in such a way that he can get the Messerschmitt in his gun-sight. The object of the operation is to align the Messerschmitt and the Spitfire's gun-sight; this process means that the pilot is part of a closed-loop system. Of course the pilot of the Messerschmitt forms part of a somewhat different control system, and will be taking evasive action that will give rise to an error as far as the Spitfire is concerned. The Spitfire pilot's task is still the same, i.e. to eliminate the error, but from the view through the gun-sight he has no objective measure of whether the error is due to his own failure to pilot the aircraft correctly or is due to the skill of the other pilot.

In pursuit tracking, by contrast, both the target position and the position of the controlled element are known relative to a fixed standard. As an example of pursuit tracking, suppose a fighter aircraft is pursuing a bomber and the movement of both aircraft is being observed on a radar screen on the ground. The radar observer is giving instructions to the fighter pilot and we assume that the instructions are being faithfully obeyed. Thus the radar observer forms part of the closed-loop that is attempting to minimise the error between the two aircraft. He can observe movements of the fighter aircraft and the bomber, each relative to a fixed standard, the ground.

Figure 52 shows an illustration of the flow of information in both pursuit and compensatory tracking. It may be seen that the

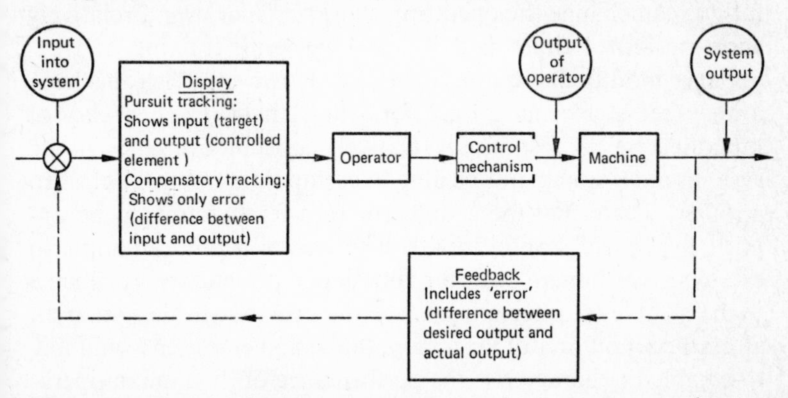

Figure 52 The flow of information in pursuit and compensatory tracking. (*Source: McCormick*, 1957)

operator forms a component in the forward loop of the system. According to Adams (1961) a subject's performance at pursuit tracking is superior to that shown during compensatory tracking, for two reasons. Firstly, since in compensatory tracking the operator cannot see the input signal, it is difficult to learn prediction responses. Secondly, the operator has no objective measure of the output, and so he has difficulty in acquiring knowledge of results. Prediction is therefore generally far superior in pursuit tracking.

As was mentioned in Chapter 4, one attempts to characterise the performance of system components by the use of the transfer function. In the case of the human operator, finding the transfer function means establishing the relationship between the sensory input and the operation of a control device. As McCormick (1957) notes:

> The intervening variables of human sensation, perception, decision, reaction time, accuracy, and strength of physical control are all imbedded within the framework of the human transfer function.

Use of the transfer function implies that a system is linear, in other words the transfer function must be equally applicable to all inputs. Since one is trying to account for terms such as decision and reaction time, one would suggest as a first guess that the human operator would be most decidedly non-linear, and experiments do indeed tend to support this. Such factors as fatigue can change the operating characteristics over a relatively short period of time; and in the long-term, training and experience modify the response patterns. However, it has also been argued (see McCormick, 1957, for details) that for a given class of input, a subject's response is relatively constant even though the systems analyst may not be able to extrapolate from one class to another. There may be a different transfer function to be applied to each of several inputs. The reason behind attempts to establish the human transfer function represents more than a wish to solve an interesting academic exercise. In designing an aircraft cockpit or a radar screen, the design engineer would like to be able to characterise the performance of the human operator in an objective way. Even with the information so far available, the limitations of the human operator are quite well known and may be accounted for in the design. To give two examples, quantitative studies of the human reaction time and psychological refractory phase have been made and semi-quantitative models can be constructed to account for delays in the system. Some tracking tasks involve the computation of velocity from position information and others require the integration of velocity information to give an estimate of future position. If the human operator is asked to perform these tasks mentally, in general he shows poor performance, and so the designer may attempt to incorporate engineering components in the system to perform such tasks and hence relieve the operator. It is also important to know the relative merits of pursuit and compensatory tracking as far as the operator's performance is concerned.

MOTION PERCEPTION

If I sit with my head and eyes held in a fixed position, and then for some reason the wall in front of me starts to move, I will readily perceive this as a movement and take some appropriate action. Let us now suppose that I keep my head fixed but move my eyes. The wall moves relative to my eyes in just the same way as before. As far as my retina is concerned the movements may well be identical, but in the second case the wall does not appear to be moving. Why?

When I move my eyes a second source of information is available to the brain apart from that derived from the retina and this information is concerned with the muscular effort involved in executing the response. When the eyes move, as opposed to the visual world, a response on the part of the extra-ocular muscles is necessary and the brain is informed of this response. When the eyes are still and motion is perceived, the brain knows that it must be the world that is moving because the information from the eye muscle system informs the brain that the eyes have not moved. In the second case the eye movement system 'tells' the brain to expect a movement of the image across the retina, and when it occurs the brain perceives this as being due to the movement of the eyes rather than being due to a movement in the visual world.

Exactly how the information is conveyed from the eye muscle system has been the subject of controversy. There are two views of how the system is constructed: the inflow theory and the outflow theory. We might also call these feedback and feedforward, respectively. The inflow theory supposes that sensory receptors are located in the eye muscles and these provide a feedback signal that informs the brain of the muscle tension and hence the position of the eye. The brain compares the information coming from the retina with the feedback signal from the muscles. This comparison enables a decision to be made as to whether the movement signal at the retina can be attributed to self-generated movement or that the visual world has moved. If the muscular tension change is appropriate to the signal coming from the retina, no perception of movement occurs.

However, according to the outflow theory it is not a feedback signal from the muscles that cancels the retinal signal, but rather it is the command signal sent to the muscles. The evidence is in favour of the outflow theory.

If one attempts to move one's own eye by pushing it with the finger the world will appear to spin round. It is easy for the reader to verify this, but be gentle. Thus, for movements that are passive as far as the eye movement controller is concerned, cancellation does not occur, which is easier to explain in terms of the outflow theory, since muscular tension would presumably be produced by the passive movement whereas no neural outflow would occur. Even stronger evidence for the outflow theory is found when we produce an after-image on the retina (see Gregory, 1966). This is done by looking at an illuminated surface for a while and by concentrating on a particular point on the surface so that a defined area of the retina is stimulated. In this way the image is fixed on the retina. We now view the after-image either in darkness or against a uniform grey surface, and again produce a passive movement of the eye. This time the image does not appear to move. If it were the case that sensors located in the muscles send a signal to the comparator, they should be stimulated by a passive movement of this kind and hence give the impression of movement. Cancellation could not occur since there is no movement of the image with respect to the retina.

The reader will recall that the essence of feedback control is that one has a check that the action has actually been executed. By contrast, with open-loop control one only knows that the command has been sent, but the system remains unaware of the consequences of the command. In the case of eye movements the inflow model involves feedback, but the outflow or feedforward model is open-loop in that the comparator judges the command that has been sent rather than the consequences of the command. If we were to prevent the eye from rotating, by some mechanical means, we would expect our perception to be disturbed if the system employed feedforward but not if the signal is fed back from the muscles. Again, the evidence supports the outflow theory in that it is possible to perceive movement without the eyes or the target moving, provided that a command has been sent to the muscles. Ernst Mach (see Gregory, 1966) prevented his eyes from moving by bunging them with putty and found that the world span round when he tried to move his eyes. The same result is obtained, perhaps less painfully, if one blocks muscular activity with curare. We can conclude that either a signal from the retina or from the eye-muscle command is a sufficient stimulus for motion perception, and these two enter a summing junction (one with a positive sign and the other with a negative) where cancellation occurs in the course of normal perception.

If the command signal were sent directly to the comparator we would expect it to arrive ahead of the retinal movement signal, since inevitably there is a delay between the time the signal gets to the muscles and the time taken for the eye to move and the image to be sent to the brain. According to Gregory the command signal is delayed so that it arrives at precisely the same time as the retinal movement signal. In systems analysis terms we would say that there is a pure delay imposed in the pathway between where the command to the muscles arises and the comparator.

Gregory raises the question as to why the system should have evolved this way. He points out that stretch receptors are present in the eye-muscles that could serve as the transducers for a feedback signal. As a possible answer Gregory suggests that a feedback system would be too slow because the feedback signal would reach the brain after the retinal movement signal. In reply one might say that the pure delay could be located in the pathway from retina to comparator, but perhaps it is a much easier task to delay the information carried in the innervation to a muscle than it is to delay retinal movement information.

According to von Holst (1954) a similar principle of comparison between muscle command and perception applies to accommodation. The phenomenon that has to be accounted for is that as a person walks towards us the image of the person on the retina may double in size but the person does not appear to get bigger. Von Holst claims that the innervation sent to the ciliary muscle increases in proportion to the increase in retinal size and the two effects cancel. In support of this, von Holst claims that drugs which affect transmission at the nerve–muscle junction of accommodation also influence our perception of size. Despite the drug study there are grounds for doubting that accommodation plays a significant role in size perception. Factors such as illumination level and pupil-size (see Toates, 1972a, for a review) appear to influence the accommodation innervation, so it would seem a poor candidate for a measuring reference. There are other more reliable cues that the brain could employ.

GROUP BEHAVIOUR

We have until now discussed control theory in so far as the individual organism is concerned, but it is important to remember that for some purposes the individual is only a subsystem within a larger system. This is perhaps particularly so in the case of insect societies, though one might be able to say the same about human social groups. As van der Kloot (1968) points out, in insect societies: 'The individual insect plays such a specialised role in the social order that the society has many of the attributes of a single, large superorganism.'

According to van der Kloot's description, in the case of temperature control the individual bee's effort is directed to maintaining the hive temperature, which normally varies only between the limits of 34·5 to 35·5 °C. If the temperature within the hive falls then the workers in the colony exert effort with their wings which produces heat and raises the temperature. Should the temperature exceed 35·5 °C, a group of workers go to the entrance of the hive and with their wings direct an air current into the hive. Another control action is taken by foragers who bring water to the hive. Workers then proceed to spread this over the surfaces of the most important parts of the hive, so that by evaporation the hive is cooled.

General Considerations and Conclusions

I hope that in the section that follows some kind of integration may be achieved between the earlier parts of the present chapter and the discussion on control theory and behaviour. For this reason we return for the moment to homeostasis.

If one accepts the theory of evolution, then it follows that mammals and birds now find themselves in an environment considerably different from that from which they evolved. In contrast to being in the sea, there is no automatically available source of fluid to bathe the cells of the body. In addition, survival depends upon being able to maintain the temperature of the internal environment. The climate in which an animal lives may well be hostile to it in so far as temperature regulation is concerned.

The temperature requirement necessitates mechanisms of heat loss and gain, and the fluid requirement means that an animal must carry around its own private sea, i.e. the plasma and interstitial fluids. The idea of homeostasis, as formulated by Claude Bernard, is that the continuance of life is dependent upon maintaining constancy of a series of bodily functions such as fluid composition and temperature. Thus the body is equipped with the thirst mechanism (involving voluntary behaviour) which forms an integral part of the body-fluid regulation loop.

In the case of temperature, reflex activities such as sweating serve to maintain constancy of the internal environment and traditionally belong in the domain of physiology. However, certain kinds of voluntary activity may also be better understood when viewed in relation to homeostasis. Thus, when I enter a cold environment I will put my coat on. This cannot be attributed to a reflex initiated by hypothalamic cooling or even by using the hypothalamus as a relay station, but none the less it ultimately serves a homeostatic end. It is in fact an example of feedforward. The sensory cold receptors convey a message to the cortex and action is taken so that the demand that is placed upon the homeostatic controller is minimised. Similarly, pain perception may almost be considered homeostatic in so far as it serves to maintain the normal state of the body by causing damaging stimuli to be avoided.

In the case of an animal such as a fish or a lizard, temperature control is purely behavioural, there being no reflex mechanisms. Thus its locomotion to a new environment is its response to a temperature stimulus. Fish appear to be equipped with temperature sensors, and these generate a signal that causes the fish to swim to a region of optimum temperature. Lizards exhibit a particularly interesting pattern of behavioural temperature regulation, which Gordon (1972) has compared to an on–off controller employing one heater. A lizard in the shade is losing heat, and the decrease in temperature causes an action analogous to 'switching a heater on' which in practice means that the lizard moves into the sun. After being in the sun for a while the temperature exceeds the 'set-point' which, to follow the analogy, causes the heater to be switched off, in other words the animal returns to the shade. The lizard may be observed to shuttle back and forth in a manner which is mathematically the same as the on–off controllers that engineers employ. In the beehive the individual bee's behaviour may be understood in terms of its contribution towards maintaining the constancy of the environment that it shares with the other bees.

For warm-blooded animals both reflex mechanisms and behavioural acts serve to maintain a controlled variable at the value dictated by some norm. Behavioural temperature regulation is the coarse control and reflex mechanisms provide the fine control. The two systems work in harmony in that conscious sensations of warm or cold arise when the skin temperature is such that core temperature is soon likely to be threatened. Room heating is an example of a coarse control, and man may form part of the loop by opening and closing windows.

The behavioural acts such as putting one's coat on are of course voluntary in so far as one can always choose not to put one's coat on. I do not want to digress into a discussion of determinism and neither do I want to stretch our terminology too far, but one might also argue that the acts are relatively involuntary in that the goal that is being aimed for, a temperature of 37 °C, is something that man has no control over. There are fixed neural circuits that determine that man will feel cold under certain conditions. So does man behave only as a negative-feedback controller when there is a fixed internal reference to be matched? We have seen that the answer is no. The nervous system employs feedback control in order to perform purely voluntary acts such as picking up a pencil.

Though we can recognise the feedback characteristics of such voluntary acts it is somewhat more difficult to draw formal control models to account for them. In order to give some meaning to the concept of the input to the system, it was argued that the brain contains some kind of model of the environment that correlates with the neural architecture of the cortex. The activity of neurons in the cortex provides the input to the controller. The act of moving a limb is guided by minimising the disparity between the actual position of the limb and the position that the cortex wills. Due to delays in the feedback pathway caused by the neural transmission time, and the time needed for computation, there is a danger that even when the man's limb reaches the desired position the part of the brain responsible for the innervation will not be told of the fact, and the limb will overshoot. The problem is answered in that the speed of response of the system appears to be sufficiently slow to guarantee stability.

A similar problem was met in the discussion of tracking behaviour. If the tracking response was at all times to be guided by feedback, the speed of response would be slow and errors would be incurred. The solution is found in employing a mixture of open-loop and closed-loop control. Tracking is closed-loop in so far as the object of the task is to eliminate error and the response is based upon error information. However, it is open-loop in that once an action has been decided upon it is then executed ballistically. This enables the speed of response to be much faster than would be possible if it were guided. Of course, the subject has to learn the responses to be associated with each error pattern and, thus, obtaining skill at tracking is a matter of learning appropriate responses largely by trial and error. We have also seen that even for pseudo-reflexes such as accommodation the response can be made on the basis of anticipation of the error.

Mittelstaedt (1957) draws attention to the basic difference between guided human responses such as slowly picking up a pencil, and essentially ballistic movements such as hammering a nail or hitting a tennis ball. Mittelstaedt's research interest is in prey capture in the mantid, and it appears that this species also employs a mixture of closed-loop and open-loop control. Thus the mantid is guided to a suitable attacking position, and then a ballistic stroke of the forelegs secures the capture of the prey. The stroke takes between 10 and 30 ms, and Mittelstaedt argues that at such a speed it could not possibly be controlled by the difference between target position and actual position. Rather the response is computed in advance and then carried out ballistically.

In lower animals ballistic movements are innate and unalterable. Prey capture is a particularly good example; after a disturbance to its visual orientation the salamander consistently makes its aim in the wrong direction and shows no learning (Sperry, 1951). By contrast, man's behaviour is characterised by its plasticity. This means that there are few innate reflexes involving man's interaction with the environment and most ballistic responses must be learned. It is on this level that feedback enters the discussion again, though in a different way to that discussed earlier, and involves the long-term modification of behaviour rather than in the short-term guidance of responses. Learning by trial and error involves feedback, since error is the difference between what is actually achieved and what should have been achieved. The word feedback has found a place in the psychologist's vocabulary, as the following quote of Fitts and Posner (1967) shows:

> Much of the incentive which motivates the activities of man comes from the consequences of his own movement. If behaviour is goal directed, then the successful approach to the goal can serve to sustain behaviour. In order to follow current usage in cybernetics, control theory and much of psychology, we shall call information arising as a consequence of the organism's response 'feedback'.

Fitts and Posner distinguish between what they call intrinsic feedback and augmented feedback. The first arises naturally, for example information from the muscles. An example of the second kind, augmented feedback, would be if the experimenter were to say: 'Yes, good. Your hand is now approaching the goal.' According to Fitts and Posner, either intrinsic or augmented feedback is able to serve the three functions of providing knowledge, motivation and reinforcement. Augmented feedback is able to improve greatly a subject's performance when tackling a task requiring skill.

Fitts and Posner also note that:

> . . . almost every act is dependent upon comparison either of feedback with input, so that he may determine the appropriateness of his previous responses, or a comparison of progress toward a goal with some conception of what is desired.

Over a period of about two decades there has been a significant change in the psychologist's view of how the nervous system organises information. The ideas of cybernetics introduced by Wiener have had a considerable influence, and this has been accompanied by exciting developments in neurophysiology. Experimental results have shown that we can no longer consider the nervous system to be reacting to environmental stimuli in a passive way, involving a one-way direction of information transmission. We know that the brain can be selective with regard to stimuli which it will attend to, and incoming information is filtered. It is anatomically possible to locate pathways going from the centre of the nervous system to the periphery; these pathways inhibit or enhance afferent information. In parallel with this research, we now know that there is not a single direction of information flow from the cortex to a motor-nerve to the muscle. Rather, messages are sent from the limbs to the areas of the brain responsible for locomotion, and the efferent signal is determined partly on the basis of information from the muscles. Nervous system activity seems to be describable only in terms of closed circuits of information transmission.

We know that our perception of certain aspects of the environment such as motion, and possibly size, arises from a comparison between an incoming sensory message and a command sent to the appropriate muscles. McFarland (1971) has borrowed this idea and extended it to cover displacement activities. To illustrate McFarland's argument, let us assume that an animal is both hungry and thirsty so that there is some motivational competition, but that at a given point in time the animal is engaged in eating. The physiological machinery involved in eating will be sending out a behavioural command and certain behavioural consequences will result. A comparator serves the purpose of comparing the behavioural consequences with those that are to be expected on the basis of the behavioural command. When the comparator signals a mismatch this is thought to lead to a switch of attention to the alternative motivation.

It was argued in the present chapter that a particularly powerful example of a command being compared with the consequences of the response is to be found in speech production. Similarly, McFarland (1971) discusses song patterns in birds:

> . . . the naive bird has a 'template' of the species-characteristic song which may, or may not, be modified by experience of the song produced by other individuals. Such modification may be restricted to a 'sensitive period', which generally occurs before the bird itself begins to sing. The extent to which the template can be modified may be limited. In the chaffinch the limitation seems to be set by the resemblance to normal song, but in other species different mechanisms operate. In the second stage of song learning there appears to be a comparison between the template and the sound produced by the bird. Singing the 'correct' song appears to be self-rewarding, for it reinforces the establishment of the song pattern in the animal's behavioural repertoire.

It is not just the active processes such as speech and song production that involve comparison with some kind of norm or template. In all probability listening to speech entails comparing what is heard with some notion of what one expects to hear. We do not, however, have a fixed template by which to compare what we hear, since this would not allow for flexibility. Rather than reacting in a predetermined way to stimuli, we create a hypothesis based partly on expectation and then check the speech pattern against this. A familiar example that seems to confirm this is as follows. As I sit here in Denmark writing this chapter, if someone now walks past the door and asks 'Hvad er klokken?', I will answer that it is twenty minutes to eleven. However, if I were to go to England and a stranger were to say the same words in exactly the same way it may well be the case that I would not understand him. In the second case I would not propose the hypothesis that the person has said 'Hvad er klokken?', and so I would search through my English vocabulary for a possible solution. For a fascinating discussion of this subject the reader can turn to Neisser (1966).

We have come a long way from the view that stimuli are reacted to like switching on a light, and it is interesting to see the principles of feedback in the context of our present view of the nervous system. Our actions seem to be initiated by something analogous to a reference value. The engineer calls the reference value the desired value of the system and our voluntary motions are initiated by desires to achieve something, for example to pick up a pencil. It appears also that perception involves a comparison of incoming information with some stored internal representation.

6 THE MATHEMATICS OF FEEDBACK CONTROL SYSTEMS

In Chapter 4 we examined how mathematical models of systems are constructed, representation being in both the time and Laplace domains. It was shown that we may sometimes reduce complex systems made up of many transfer functions to a representation involving only one transfer function. In Chapter 5 we considered in a descriptive way a special class of system, the negative-feedback system. The aim of the present chapter is to go some way towards integrating Chapters 4 and 5, in other words, to show how the tools of mathematical analysis may be applied to feedback systems. The negative-feedback systems that we will analyse are made up from the basic components that we are already familiar with from Chapter 4.

The mathematical examples presented in this chapter all give clear, simple and quite unambiguous results; this is rarely the case with biological systems, and in itself may cause the reader to doubt the relevance of such techniques. However, science proceeds by a series of approximations to the truth, so perhaps we ought not to feel too bad about having qualitative descriptions of biological control systems accompanied sometimes by a mathematical model that in many cases provides a very good first approximation.

We can consider first the equation that defines the performance of a negative-feedback controller having linear components that may be characterised by transfer functions. This is shown in Figure 53(a). $G(s)$ is the forward-loop transfer function and therefore describes the operation relating $\bar{e}'(s)$ (not to be confused with the exponential e) to the output of the control system $\bar{y}(s)$. $H(s)$ is the feedback loop transfer function. A mathematical description of the input, that is to say the Laplace transform of the input is represented by $\bar{x}(s)$. The Laplace trans-

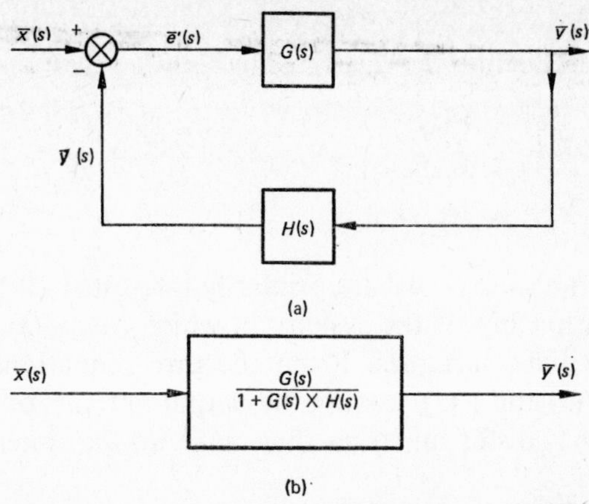

Figure 53 The reduction of a negative feedback system by the use of the transfer function. The system of (a) may be reduced to the system shown in (b).

form of the output $\bar{y}(s)$ is multiplied by the transfer function $H(s)$ to give $\bar{y}'(s)$. $\bar{y}'(s)$ is compared with $\bar{x}(s)$ in order to give a measure of the error that in Figure 53(a) is called $\bar{e}'(s)$. The circle with a cross in it is a summing junction, at which input $\bar{x}(s)$ has $\bar{y}'(s)$ subtracted from it. In the present case we are finding the error between the input and a measure of the output. The use of the expression 'measure of' the error and 'measure of' the output to describe $\bar{e}'(s)$ and $\bar{y}'(s)$ was deliberate. If error is defined as input minus output, then in the terms we have just used this is given by $\bar{x}(s)-\bar{y}(s)$, whereas $\bar{e}'(s)$ is in fact given by $\bar{x}(s)-\bar{y}'(s)$. It perhaps does not matter too much whether we call error the difference between input and output or the signal $\bar{e}'(s)$, so long as we know what we are talking about. The signal $\bar{e}'(s)$ is the same thing as the error between input and output (which is denoted by $\bar{e}(s)$) only when $\bar{y}'(s)$ is the same as $\bar{y}(s)$ – in other words when the effect of $H(s)$ is to give a gain of 1. In fact $H(s)$ often has the value of 1, but it is important to consider the general case. For example, there may be a delay in the feedback pathway such that $\bar{y}'(s)$ is a faithful measure of $\bar{y}(s)$ in the steady-state but not in the time before the steady-state is reached. If $H(s)$ does represent a delay element or a gain having a value other than unity, this can have a most profound effect on the performance of the system, changing it considerably from what would be the case if it were unity.

By inspection of Figure 53(a) we are able to write down three equations that define the relationship between the variables, i.e.

$$\bar{y}(s) = G(s) \times \bar{e}'(s)$$
$$\bar{e}'(s) = \bar{x}(s) - \bar{y}'(s)$$
$$\bar{y}'(s) = H(s) \times \bar{y}(s)$$

Often, but not always, we are primarily interested in the input–output relationship of the system, in which case $\bar{e}'(s)$ and $\bar{y}'(s)$ are intermediate variables. From the three equations we can eliminate $\bar{e}'(s)$ and $\bar{y}'(s)$, leaving the output in terms of the input and the two transfer functions that make up the system.

$$\bar{y}(s) = G(s) \times \bar{e}'(s)$$
$$\bar{y}(s) = G(s) \times [\bar{x}(s) - \bar{y}'(s)]$$

but since $\bar{y}'(s) = H(s) \times \bar{y}(s)$

$$\bar{y}(s) = G(s) \times [\bar{x}(s) - H(s) \times \bar{y}(s)]$$
$$\bar{y}(s) = G(s) \times \bar{x}(s) - G(s) \times H(s) \times \bar{y}(s)$$
$$\bar{y}(s) + G(s) \times H(s) \times \bar{y}(s) = G(s) \times \bar{x}(s)$$
$$\bar{y}(s) = \frac{G(s)}{1 + [G(s) \times H(s)]} \times \bar{x}(s)$$

This has established the input–output relationship of the system in terms of a transfer function, and as Figure 53 shows, we may replace the negative-feedback system of (a) by the single equivalent operation shown in (b). We are now able to predict the output that will occur in response to a specified input if we know the components $G(s)$ and $H(s)$. The use of the transfer function method is best illustrated by examples.

Example 13

A negative-feedback control system contains an integrator in the forward loop and a feedback pathway of gain 1. What is the response to a unit step input? The system is shown in Figure 54(a). In Figure 54(b) it has been translated into a transfer function notation, and the integrator therefore becomes $1/s$. $H(s)$ has a value of 1. $G(s) = 1/s$, $H(s) = 1$, and $\bar{x}(s) = 1/s$.

$$\bar{y}(s) = \frac{G(s)}{1+[G(s)\times H(s)]}\times \bar{x}(s)$$

$$= \frac{1/s}{1+1/s\times 1}\times 1/s$$

$$= \frac{1}{s+1}\times 1/s$$

Breaking this up into partial fractions

$$\frac{1}{s(s+1)} = A/s+B/(s+1)$$

$$1 = A(s+1)+Bs$$

$$1 = As+A+Bs$$

$$A = 1$$

$$B = -1$$

$$1/[s(s+1)] = 1/s-1/(s+1)$$

Therefore $\bar{y}(s) = 1/s-1/(s+1)$

after looking at the inverse Laplace transform table in Chapter 3 we know that

$$y(t) = 1-\exp(-t)$$

The response may therefore be broken down into two parts: a step of height 1 and an exponential; these are shown in Figure 54(c). To calculate the value of the decaying exponential at time zero we make t equal to zero.
Thus

$$\exp(-t) = 1/\exp(t)$$

and when $t = 0$

$$1/\exp(t) = 1/\exp(0) = 1/1 = 1$$

when $t = \infty$

$$1/\exp(t) = 1/\exp(\infty) = 1/\infty = 0$$

The term $\exp(-t)$ decays from a value of 1 at time zero to a value of zero at time infinity. The output $y(t)$ is given by the unit step minus $\exp(-t)$.

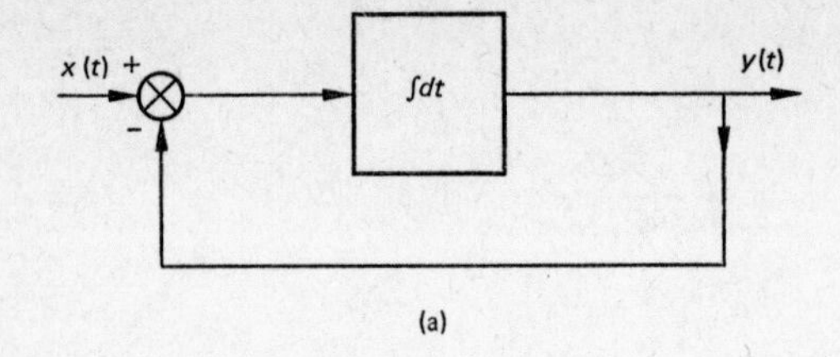

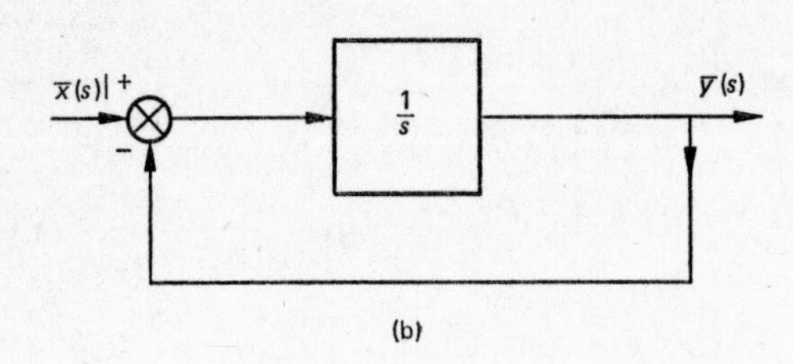

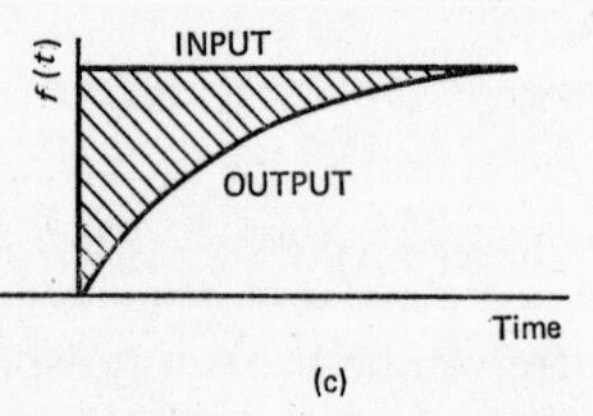

Figure 54 An integral controller. (a) In terms of the time variable, (b) in terms of Laplace, (c) the response to a unit step input.

Example 14

What is the response to a unit impulse input for the system described in example 1?

In this case the Laplace transform of the input is $\bar{x}(s) = 1$. $G(s) = 1/s$, and $H(s) = 1$.

$$\bar{y}(s) = \frac{G(s)}{1+[G(s)\times H(s)]}\times\bar{x}(s)$$

$$= \frac{1/s}{1+1/s\times 1}\times 1$$

$$= 1/(s+1)$$

$$y(t) = \exp(-t)$$

Example 15

A negative feedback system has unity feedback. For a unit step input, the output is a step of height 0·9. What is the forward loop transfer function? $x(t) = 1$, and $y(t) = 0{\cdot}9$.

$$\bar{x}(s) = 1/s,\ \bar{y}(s) = 0{\cdot}9/s,\ H(s) = 1$$

$$0{\cdot}9/s = \frac{G(s)}{1+[G(s)\times 1]}\times\frac{1}{s}$$

$$0{\cdot}9+0{\cdot}9\times G(s) = G(s)$$

$$0{\cdot}9 = 0{\cdot}1\times G(s)$$

$$G(s) = 9$$

Steady-State Errors

This subject was discussed in the last chapter in the context of heating a room. It was shown that if the room has perfect insulation – in other words it acts as a pure integrator as far as heat production is concerned – then the output is able to match the input perfectly. However, if in the steady-state it is necessary to apply heat to the room to maintain its temperature and the heat production is proportional to the error, then output will not be in perfect agreement with the input. There will be a steady-state error upon which the heating effect is dependent. We are now in a position to re-examine this problem, but from a mathematical point of view.

The steady-state refers to what is happening when all of the transient terms die out. In Example 13, $\exp(-t)$ is the transient term which becomes zero as t approaches infinity. We are left with the steady-state solution $y(t) = 1$, and since the input $x(t) = 1$, the steady-state error (the difference between input and output in the steady-state) is zero. In this case we have been dealing with an integral controller, i.e. a system that contains an integrator in the forward loop. It is because of the integrator that the system reduces the error to zero.

Though the steady-state error is zero, there is an initial transient error. In Figure 54(c) the shaded area is the integral of error with respect to time. The vertical distance between input and output is the error and the horizontal distance is time. Since output is proportional to the integral of error, then obviously there must be an initial transient error in order for there to be an output. While there is an error the output will be an increasing quantity, but when the output is steady then the error must be zero. To take a simple analogy, in order to have a volume of water in a bucket there must have been a flow of water into the bucket. However, when the volume reaches a constant value then the flow must be zero. If we make the flow into the bucket dependent upon the difference between the level of water in the bucket and some measuring reference, then the volume will increase until the height of water matches the reference, at which point the flow will be switched off.

To return to the control system, if instead of the integrator, a pure gain K were to be located between the error detector and the output, the system would be a proportional controller. That is to say output is proportional to the error. We can investigate the effect of applying the same input as in Example 13, but changing the system from integral $[G(s) = 1/s]$ to proportional $[G(s) = K]$.

$$\bar{y}(s) = \frac{G(s)}{1+[G(s)\times H(s)]} \times \bar{x}(s)$$

$$= \frac{K}{1+K} \times \frac{1}{s}$$

$$y(t) = \frac{K}{1+K}$$

It can be seen that for an input of 1 in this system, we never obtain an output of 1, but rather an output of $K/(1+K)$. Since the input is 1, the error is at all times given by

$$1 - K/(1+K) = 1/(1+K)$$

In other words a steady-state error is retained. As K is made larger, i.e. we increase the forward loop gain, then so the steady-state error gets smaller, but it will not be eliminated completely since only an integral controller can achieve that. Since a proportional controller is one in which output is proportional to error, then obviously an output cannot exist without an error. In Example 15 we met a system that, after applying a step of height 1 to it, yielded a step of height 0·9 at the output. The steady-state error in this case is 0·1. Error (0·1) is multiplied by the forward-loop gain (9) in order to give the output (0·9). The properties of the system are such that one could not expect to get 1 out for putting 1 in.

This is probably an opportune moment to draw attention to a very common misunderstanding that students encounter when beginning a study of feedback systems. If the student is presented with a feedback system having a forward-loop gain of 9 and an input of 1, he or she has an inclination to answer that the output is 9. The input appears to be multiplied by the forward-loop gain in order to give the output. In fact, the forward-loop gain does not 'see' the input 1·0, but rather the error 0·1. The object of a negative-feedback system is to reproduce the input (not to amplify it) and in order to do this it has to amplify the error.

So much for the response of systems to a step input. We will now return to the integral controller, and examine what happens when we apply a ramp input $x(t) = t$ to it.

$$\bar{x}(s) = 1/s^2 \qquad \text{since} \quad x(t) = t$$

$$\bar{y}(s) = \frac{G(s)}{1+[G(s) \times H(s)]} \times \bar{x}(s)$$

$$= \frac{1/s}{1+1/s \times 1} \times 1/s^2$$

$$= \frac{1}{(s+1) \times s^2}$$

breaking up by means of partial fractions

$$\frac{1}{s^2(s+1)} = A/s^2 + B/s + C/(s+1)$$

$$1 = A(s+1) + Bs(s+1) + Cs^2$$

$$1 = As + A + Bs^2 + Bs + Cs^2$$

$$1 = A$$

$$B = -1$$

$$C = 1$$

$$\bar{y}(s) = 1/s^2 - 1/s + 1/(s+1)$$

$$y(t) = t - 1 + \exp(-t)$$

When t is infinity, $\exp(-t)$ is zero, so $y(t)$ is given by $t-1$. When t is zero, $y(t)$ is given by $0-1+\exp(-0)$ which equals zero. Input and output are as shown in Figure 55. It may be seen

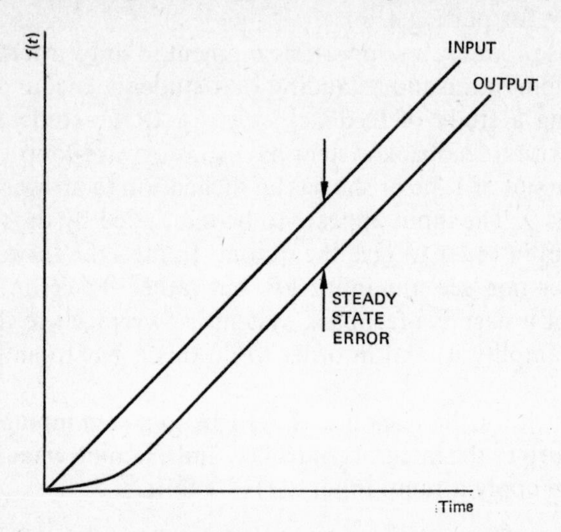

Figure 55 The response to a ramp input, showing the presence of a steady-state error.

that in the steady-state the output has the same velocity, i.e. slope or rate of change, as the input, but lags behind it by a constant amount. Thus, the same integral controller that exhibited no steady-state error for a step input has a constant steady-state error in response to a ramp input. The reason is as follows.

The forward-loop transfer function $1/s$ is that of an integrator, the input of which is error. Since the integral of a constant K_1 is a ramp K_2t, in order to get an output K_2t the integrator must 'see' a constant K_1. Since the input to the integrator (not to be confused with $x(t)$, the input to the control system) is error this must be constant, in other words output from the controller [$y(t)$] must lag behind the input [$x(t)$] by a constant amount.

The analogy of a bucket of water with a hole in the bottom may make the problem somewhat clearer. If we pour water into a bucket (the volume that we pour being defined by K_1t), water will ultimately pour out of the hole at the bottom at the same rate at which we pour it in. However, a certain volume will be retained in the bucket, since the flow is dependent upon the presence of a certain head of water in the bucket. The volume leaving the bucket is given by the volume poured in (K_1t) minus the amount left in the bucket.

A steady-state will be reached where the volume in the bucket is constant, flow out being equal to flow in. Volume poured in may be compared to the input to the controller, volume leaving is then the analogue of the output and we may compare the error to the volume retained in the bucket.

Speed of Response

Let us for the moment consider Example 13 but with a forward-loop transfer function K/s, where K is a constant. The system is still an integral controller but in addition there is a gain term associated with the forward loop. If the reader cares to repeat the calculation using the same input, the result $y(t) = 1 - \exp(-Kt)$ will be obtained. The first point to note is that when t approaches infinity $\exp(-Kt)$ approaches zero, and so there will be no steady-state error whatever value K takes. The forward-loop gain term has no influence on the steady-state performance. However since the transient term is given by $\exp(-Kt)$, K determines the speed of response. This is shown in Figure 56.

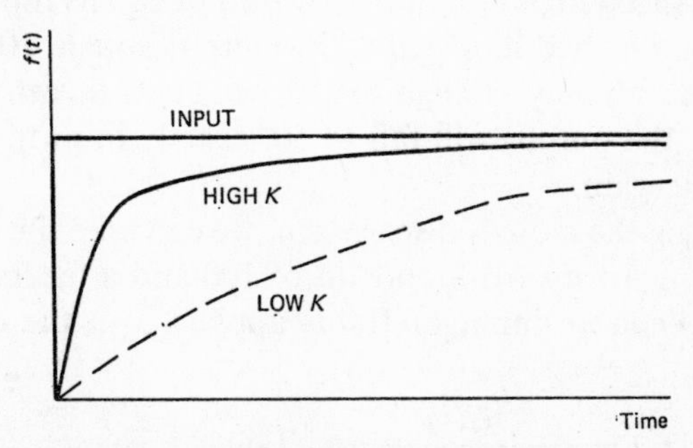

Figure 56 The continuous curve shows an exponential while the broken curve shows a slightly slower exponential. The final position reached is the same in each case. K is slightly higher for the faster exponential.

If K is small, the exponential decays slowly, in other words the response rises slowly. A large value of K means that

$\exp(-Kt)$ rapidly disappears, and therefore the output rises very rapidly.

If we have a bucket of water, and a flow into the bucket is caused by the distance between the level of water and some reference level, such that flow $=$ (distance) $\times$ (K), then water will flow into the bucket until the reference level is reached, whatever value of K is chosen. In other words, there can never be a steady-state error. However, K determines how fast the reference value is reached. If we increase K then the flow is faster for any given distance from the reference value.

In Example 13 we considered the case of $1/s$ which is just a particular case of K/s when K equals 1.

Gain Changes

In Chapter 5 we compared open-loop and closed-loop heating. It was mentioned that open-loop systems are very sensitive to changes in the characteristics of their components whereas closed-loop systems are relatively insensitive. One change that can occur in a control system is that the gain of the forward loop can alter, and it is therefore essential to know what effect this is likely to have on the output.

To take the open-loop situation first. An open-loop system has an input of 100 and a forward-loop gain of 1. Since the equation of the system is simply that output equals input multiplied by the forward-loop gain, the output equals 100. If the gain should suddenly change by 50 per cent, in other words become 0·5, the output will fall to 50, exactly in proportion to the change in gain.

Considering the closed-loop system, we can take the case of a system having a forward-loop gain of 100 and a feedback-loop gain of 1. When an input of 100 is applied to it the output is given by

$$\bar{y}(s) = \frac{G(s)}{1+G(s)\times H(s)} \times \bar{x}(s)$$

$$= \frac{100}{1+100} \times \frac{100}{s}$$

$$y(t) = \frac{10000}{101} \simeq 99$$

If the gain should now change by 50 per cent to become 50, the output becomes

$$\bar{y}(s) = \frac{50}{1+50} \times \frac{100}{s}$$

$$= \frac{50}{51} \times \frac{100}{s}$$

$$y(t) \simeq 98$$

A change of gain of 50 per cent produces a change in output of 1 per cent, as opposed to a 50 per cent change in the case of an open-loop system. This is one of the reasons why engineers employ negative-feedback systems, since it is not an important requirement to have constant characteristics of the forward-loop components of the system. By contrast, in the case of open-loop systems a change in the forward-loop gain has a drastic effect on the output. As was shown in the last section, changing the gain associated with the forward loop of an integral controller has no effect on the steady-state response to a step input, though it changes the speed of response. For the proportional controller under similar circumstances a change does occur in the steady-state value attained, although it is a small change.

Disturbance in the Forward Loop

Figure 57(a) shows that an open-loop system having an input $\bar{x}(s)$ and an output $\bar{y}(s)$ is normally given by $G(s)$ multiplied by $\bar{x}(s)$, but then a disturbance $\bar{d}(s)$ is somehow added to the forward loop, and now the output becomes

$$\bar{y}(s) = G(s) \times \bar{x}(s) + \bar{d}(s)$$

In other words the disturbance is transmitted to the output without any attenuation. This situation should be contrasted with the closed-loop situation shown in Figure 57(b), where a disturbance appears at the output end (the location of the disturbance is important).

Since the system is linear we can calculate the output as being the linear sum of the response to the input when the disturbance

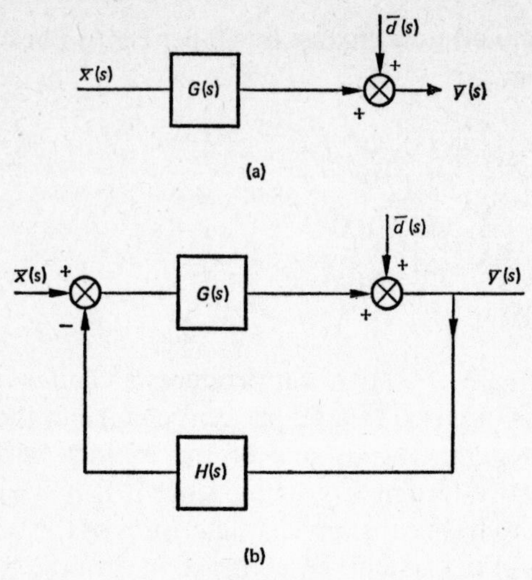

Figure 57 The point of interaction of a disturbance d(s). (a) In an open-loop system, (b) in a closed-loop system.

is zero plus the response to the disturbance when the input is zero.

The response to $\bar{x}(s)$ that we can call $\bar{y}_1(s)$ will be given by

$$\frac{G(s)}{1+[G \times H(s)]} \times \bar{x}(s)$$

The response to the disturbance $\bar{d}(s)$ is not so easy to visualise but the best way to do the calculation is to imagine a forward loop that now consists of a straight connection between $\bar{d}(s)$ and $\bar{y}(s)$, in other words, a gain of 1. This is of course not the forward loop that we have been used to because we have normally had $\bar{x}(s)$ as the input. However, for the purposes of calculating the output in response to $\bar{d}(s)$, we have assumed $\bar{x}(s)$ to be zero. Our forward loop is that which links the input, now $\bar{d}(s)$, to the output $\bar{y}_2(s)$. We also have a different feedback loop as far as $\bar{d}(s)$ is concerned. The signal is fed back via $G(s)$ in order to reach the point of the disturbance $\bar{d}(s)$. Therefore, the equation that defines the output in response to the disturbance $\bar{d}(s)$ is

$$\bar{y}_2(s) = \frac{1}{1+[G \times H(s)]} \times \bar{d}(s)$$

The total output $\bar{y}(s)$ will be made up of the sum of the components $\bar{y}_1(s)$ and $\bar{y}_2(s)$. This we are able to do by means of the principle of superposition.

$$\bar{y}(s) = \frac{G(s)}{1+[G \times H(s)]} \times \bar{x}(s) + \frac{1}{1+[G \times H(s)]} \times \bar{d}(s)$$

Let us take as an example the case when $G(s)$ is a pure gain of 100 and $H(s)$ is a gain of 1. The output will be held close to the input, since the ratio of output to input is 100/101, but on the other hand the disturbance input to the forward loop [$\bar{d}(s)$] will be multiplied by 1/101. In other words, almost all of the input will reach the output, but only a tiny fraction of the disturbance will be transmitted. The reason for this is that the disturbance sets up a corrective action that tends to eliminate it as soon as it appears.

The open-loop system was able to do absolutely nothing to correct for the forward-loop disturbance, whereas the feedback system reduced it to a tiny fraction of its original value. This example demonstrates the essence of what feedback control is about, i.e. the ability of a system to maintain an output in the face of a disturbance.

The Analogue Computer as a Tool for Analysis

Electronic computers are basically of two kinds: digital and analogue. The digital computer will be discussed in Chapter 8 in the context of the simulation of behavioural aspects of homeostasis. This section comprises a discussion of the analogue computer.

Before we turn to computing, this is probably a good place to stop and take stock of what has been said so far. In this way the possibilities for using the computer will fit into the general context of control systems analysis.

In the discussion of mathematical models it was shown that components may be either linear or non-linear. If the components are linear then we may employ the Laplace transform method, where operators take the form of transfer functions. The time-dependent functions are converted into Laplace transforms, and the output for a particular input may be calculated.

If the system is almost linear then we can form a linear approximation and apply linear techniques. This method gives a first rough estimate of the response.

Essentially, the transfer function method involved translating from one domain (time) that we are familiar with into a new domain (Laplace). We lose nothing by this transformation, and indeed finally translate the problem back into the time domain. As useful as the Laplace transformation is, it unfortunately applies only to linear systems, and it would be quite wrong to try to stretch its use to systems containing definite non-linearities, such as thresholds. There is nothing wrong with including non-linearities on a block diagram in order to show one's theory of how a system works, but it must be understood that an analytical solution by means of Laplace cannot be obtained. What we need is a model of the system in a language that can account for the non-linearities. The computer provides us with such a language.

When we construct a block diagram and use it as an analytical tool, the terms that we use are only analogous to the real system. K and Kt are not the same as flow but rather provide us with an unambiguous description of flow. When we say a step input of flow rate and characterise this by the constant K, the term K refers to a rate of water flow. When we further transform K into K/s by means of Laplace we have changed it yet one stage further from what we can observe. However, what is important is that there should be a unique correspondence between these transformations, so that we lose no information in going from actual water flow to K and then to K/s. It is indeed the case that no information is lost.

Let us consider another possible transformation, from the real system to an electronic model that embodies the structure of the real system. In normal calculations the number 10 can refer to 10 cm^3 of water, but in an electronic model a volume of 10 cm^3 might be represented by an electrical voltage of 10 volts. If we had an electronic multiplier that could multiply this voltage by 2 in order to produce 20 volts, then this would correspond to 20 cm^3 in the real system. An analogue computer model of a system is really nothing more than such a translation from the real system. It is possible to compare every variable that takes the form of a voltage in the electronic model with a corresponding variable in the real system. The electronic components or hardware in the model perform operations such as multiplication by a constant or integration on the voltages that they receive, and thus their performance characteristics may be compared to the transfer function of a block diagram, except that there is no restriction of linearity.

Figure 58(a) shows one of the basic components of an analogue computer: the summing unit. In the present example there

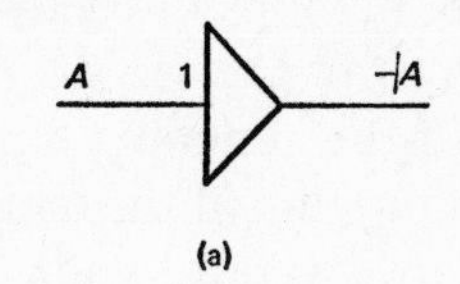

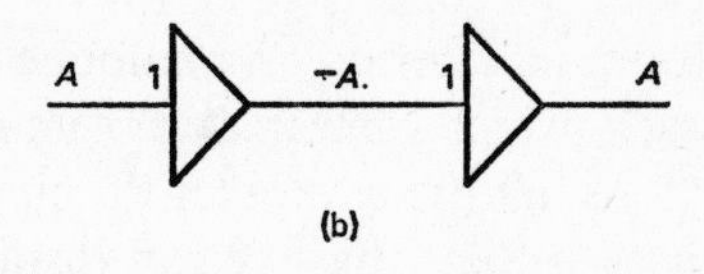

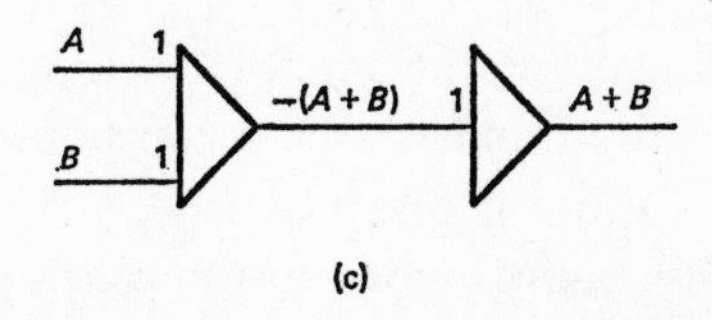

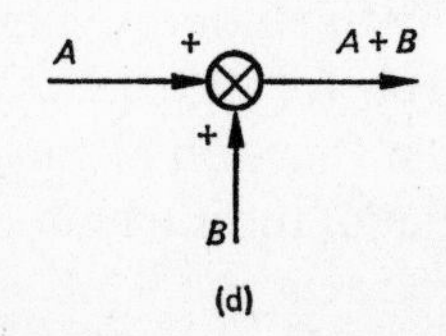

Figure 58 (*a*) *A summing unit*, (*b*) *two summing units in series so that the two sign changes cancel*, (*c*) *these two units form the sum* $A+B$ *without a sign change*, (*d*) *a block diagram equivalent of* (*c*).

is just one input A which gives an output $-A$, since the input goes into the terminal marked 1 this means that the gain is 1. The reason that the output is $-A$ and not A is that electronic amplifiers inevitably produce a change in sign of the voltage that they receive. A is a voltage, for example 10 volts, and results in an output of -10 volts. In Figure 58(b) the first amplifier changes the sign from A to $-A$, and the second amplifier gives another sign change so that we get back to A again.

Figure 58(c) shows two inputs A and B. The summing unit adds these to give $A+B$, but introduces a sign change so that we get $-(A+B)$. The output of the second amplifier is $A+B$. As far as we are concerned the summing amplifier is a black box giving an output that is the sum of the input voltages, and introducing a sign change. If we want to know the sum of two terms A and B, all we need to do is to apply voltages proportional to A and B at the input to the first amplifier, and then measure the output of the second amplifier which will be proportional to $A+B$. This electronic unit forms an analogue or model of the summation process, in which we translate from numbers or from time varying functions, let the computer do our calculations for us, get the result as a voltage that we can display and relate to the real system. The two electronic amplifiers together are equivalent to the summing unit shown in block diagram form in Figure 58(d).

Another component of the analogue computer is the integrator, and this is represented as the component on the left in Figure 59(a). The output is given by $-\int_0^t A\,dt$ which is then inverted. This process is analogous to the block diagrams shown in Figure 59(b) and (c). If we want to calculate the integral of a function we make the function take the form of a voltage, apply this to the integrator and measure the output which will then give us the integral of the function. A summing unit follows the integrator and this corrects for the sign change. For example, to find the integral of a step function at time zero a constant voltage is applied to the input of the integrator, and then the voltage at the output is measured.

In Figure 59(d) it may be seen that there is an additional feature of the integration process marked I.C. that was not included in Figure 59(a) and which stands for initial conditions.

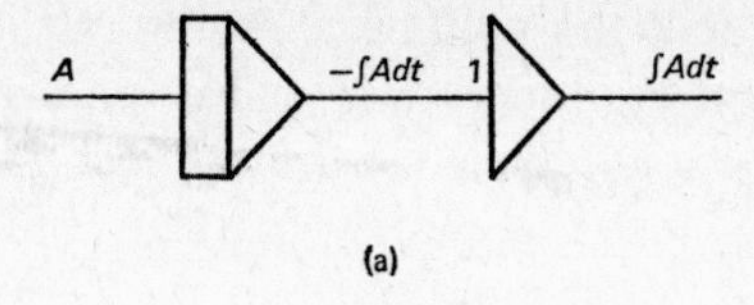

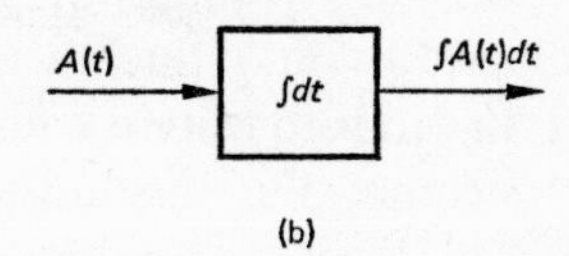

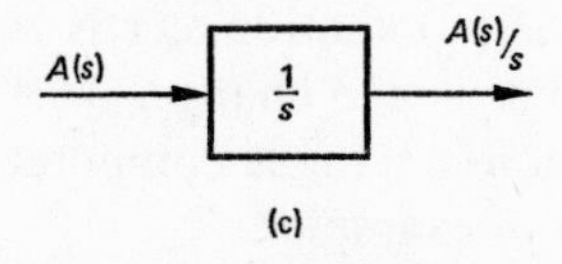

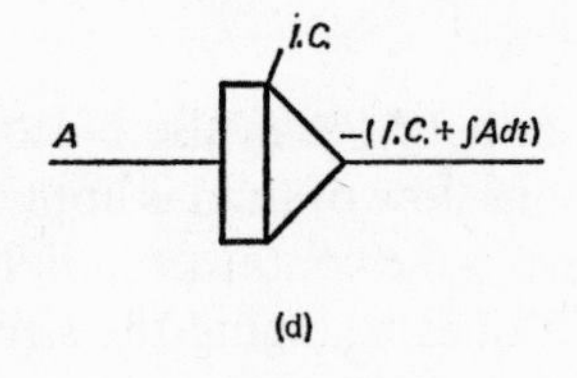

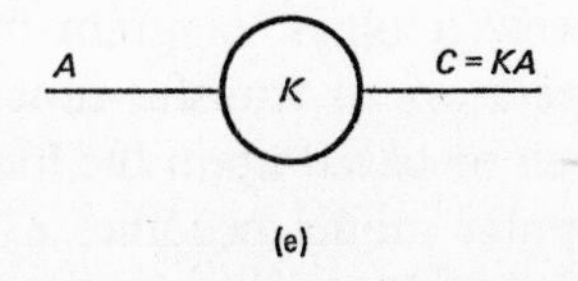

Figure 59 *Analogue computer components and their equivalents.* (*a*) *An integrator followed by a sign change performs the operation of integration with no net sign change,* (*b*) *the equivalent operation in a block diagram and time representation,* (*c*) *the equivalent operation in Laplace-transfer function representation,* (*d*) *an integrator including initial conditions,* (*e*) *a potentiometer.*

If we wanted to make an analogue computer model of the volume–flow relationship of a bucket of water, then flow would form the input to the integrator and volume the output. Initial conditions represent the volume of water in the bucket at the start of the period of observation: call it time zero. In practice, when working with analogue computers each integrator has its input and output terminal clearly marked, and there is also a site for applying initial conditions in the form of an electric voltage.

Normally the computer has voltages of plus and minus 100 volts (on some machines 10 volts). Such a voltage is known as a machine unit (MU). In order to obtain voltages less than those supplied from the machine, the machine voltage is passed through a potentiometer. The potentiometer is a component that gives a gain of between 0 and 1. Figure 59(e) shows an input A and an output C, where $C = AK$, and K is a constant between 0 and 1·0. The potentiometer is a passive component and does not introduce a sign change.

The use of the various analogue computer components is best illustrated by means of examples.

Example 16

A container has an outlet hole in the bottom and is empty at time zero. A continuous flow of fluid is applied to the container. What does the volume in the container look like as a function of time, and what effect does changing the size of the outlet hole have on the volume?

Problems may either be translated directly into an analogue model, or alternatively a block diagram may be constructed using either time operators or transfer functions, and the analogue model may then be based upon the block diagram. In the latter case the computer model becomes a model of a model, but no information is lost in such transformations since there is a direct correspondence between the time domain, the Laplace domain, and the analogue model (unless of course the translator makes a mistake).

Figure 60(a) shows a block diagram of the system. The transform of volume [$\bar{v}(s)$] is related to the transform of flow [$\bar{f}_i - \bar{f}_o(s)$] by the transfer function $1/s$, in other words the operation of integration. Net flow is given by flow in [$\bar{f}_i(s)$] minus flow out [$\bar{f}_o(s)$], and we assume that flow out is propor-

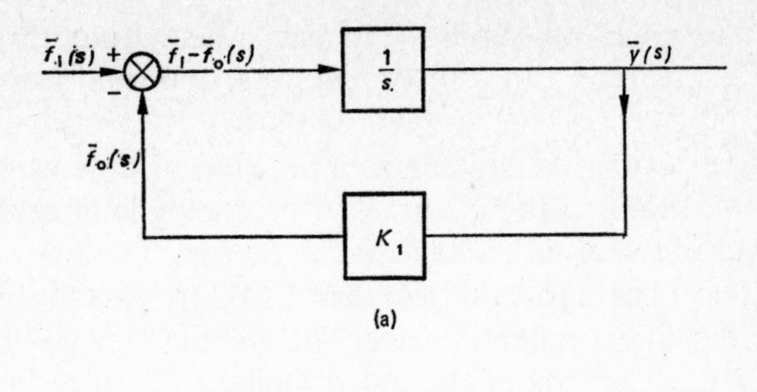

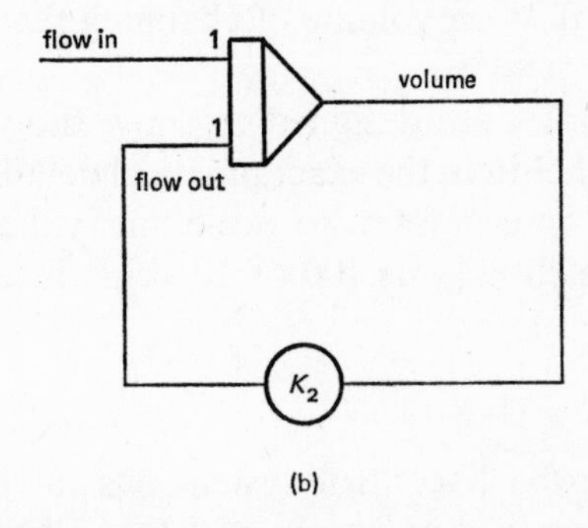

Figure 60 (a) Block diagram, and (b) analogue computer models of the flow of water from a container with an outlet hole in the bottom. Flow out is proportional to volume. Volume is the integral of the net flow in.

tional to volume. The gain K_1 relates volume to the flow out of the container, the value of K_1 depending upon the size of the outlet hole.

In Figure 60(b) an analogue model of the system is presented. K_2 is the value of a potentiometer gain that relates flow out to volume. Volume is the output of the integrator unit, and flow in is a voltage applied to the input of the integrator. Since the container is empty at time zero, I.C. is zero volts. It should be noted that though we are simulating quantities having different units such as cm^3 and cm^3/min, in the simulation all variables have volts as the unit, except, of course, time. If there were a volume at time zero then a voltage proportional to this volume would have been applied. Flow out of the container also forms an input to the integrator, but of course needs to have the opposite sign to the flow in. This will result without using any additional components. Since the integrator itself introduces a sign change we can simply feed back the output through a potentiometer and this will provide the opposite sign to the voltage representing the flow in.

In order to observe volume as a function of time we apply a constant voltage to the integrator (the analogue of a constant flow) and measure the voltage at the output. To obtain a constant voltage having a value less than 1 MU we pass the machine voltage through a potentiometer. The volume may be observed using various settings of the potentiometer K_2, in order to investigate the effect on volume of changing the size of the outlet hole.

These examples are designed to show the principles behind simulation rather than the exact practical details. When building the model it is necessary to scale the voltages so that the maximum (which may be 100 or 10 volts) is never exceeded.

Example 17

A negative-feedback control system has an integrator in the forward loop as well as a gain of 10, and has unity feedback. What effect does halving the forward-loop gain have on the steady-state error that occurs in response to a ramp input?

Figure 61(a) shows the system in block diagram form, which in Figure 61(b) has been translated into an analogue computer model. An amplifier is employed in order to obtain a gain of 10. Normally these amplifiers have inputs marked 1 and 10, and we use the 10 input to provide a gain of 10. In order to get a gain of 10 the potentiometer K_2 will of course have to be at its maximum setting of 1·0. An integrator follows the potentiometer – the output of this component being the output of the system. Since there are two sign changes in the forward loop, their effect will cancel, and therefore the output will have the same sign as the input. If we were simply to feed back the output to the summing amplifier this would produce positive feedback. Thus a sign change must be introduced into the feedback pathway. This is done by passing the signal through summing amplifier 2, using the 1 input to give a gain of -1.

It should be noted that the summing amplifier differs from the summing junction of the block diagram in that only sum-

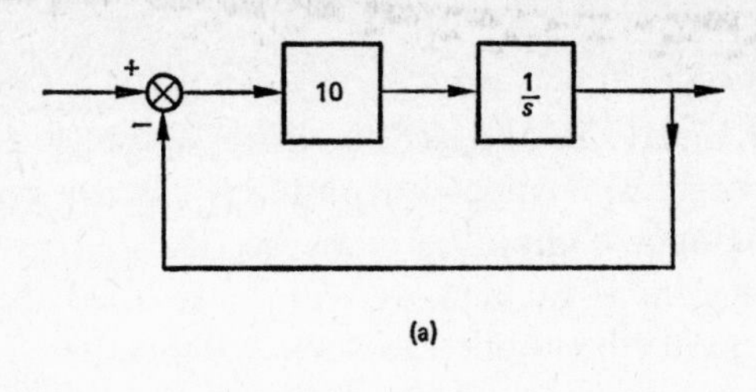

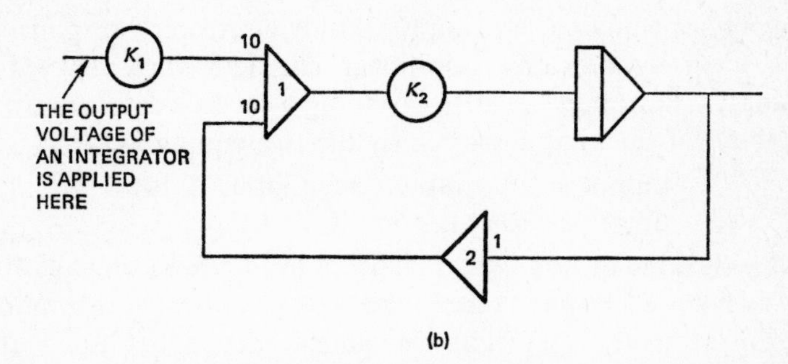

Figure 61 (*a*) *Block diagram, and* (*b*) *analogue computer models of a feedback control system having an integrator in the forward loop.*

mation can occur at the summing amplifier, whereas signals can be both added and subtracted at a summing junction. This means that in order to perform subtraction on the analogue computer the voltage must be made negative by means of a sign change before it reaches the summing amplifier.

How do we obtain a ramp that can be applied to the input? Since we are only provided with constant voltages (i.e. 1 MU) on the computer, the ramp is obtained by means of integrating a constant voltage. In Figure 61 potentiometer K_1 determines the magnitude of the voltage applied to the integrator and therefore determines the slope of the ramp.

To investigate the effect on the steady-state error of changing the gain, we measure the voltage that corresponds to the error. This is the output of summing amplifier 1 (actually this is 10 times the error since the inputs go into the 10 position, but the factor 10 will be constant) and is measured by applying a voltmeter or some automatic recording apparatus at this point. The voltage is recorded with the potentiometer setting at both 1·0 and 0·5, in order to give gains of 10 and 5.

NON-LINEARITIES

Examples 16 and 17 both described linear systems. Though it is sometimes very convenient to simulate linear systems, the power of analogue computing really becomes apparent when a non-linear system is tackled. It will be recalled that Laplace transforms cannot be applied to such systems.

Since the computer forms an electronic model of the system that we are studying, we simply employ electronic components that themselves exhibit non-linear characteristics. Take for example a hard non-linearity (meaning a severe departure from linearity) of the kind described by the two equations

$$\text{output} = 1{\cdot}0 \times \text{input (when input} \leqslant 10)$$
$$\text{output} = 10 \text{ (when input} > 10)$$

This is the kind of non-linearity shown in Figure 38 on page 102 and consists of two regions: a linear slope and a flat saturation region. It is the fact that the gain is dependent upon the input that precludes the employment of conventional mathematics. In the analogue computer model the gain is made to be 1 when the input is less than 10 volts by feeding the input into the 1 position of the summing amplifier. In order to simulate the saturation region, electronic components are connected to the output of the amplifier in a manner that prevents its voltage from exceeding 10 volts. This is a simple procedure involving only a constant voltage source and a diode. As far as we are concerned we are dealing with a black box whose output is a linear function of the input when this is below 10 volts, but when the input voltage exceeds 10 volts the output remains at 10 volts.

Another non-linear feature of the computer is its ability to multiply two variables together. Thus a variable x may be fed into one input of a multiplier unit, a variable y into the second and the product xy forms the output. A simple adaptation of the multiplier unit enables division to be performed.

We are now in a position to consider the application of the computer to a problem in biology that involves a non-linearity.

AN ANALOGUE COMPUTER MODEL OF DRINKING

McFarland (1971) presented a block diagram model and analogue computer model of drinking, the block diagram being shown in Figure 62(a). It is a simplified version of the real system, but none the less serves to show its essential features.

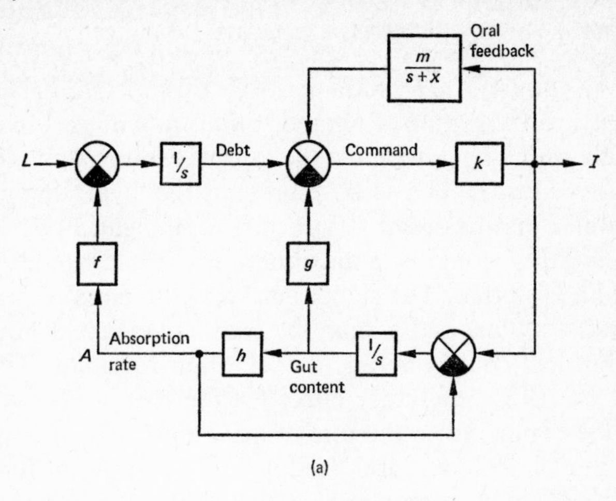

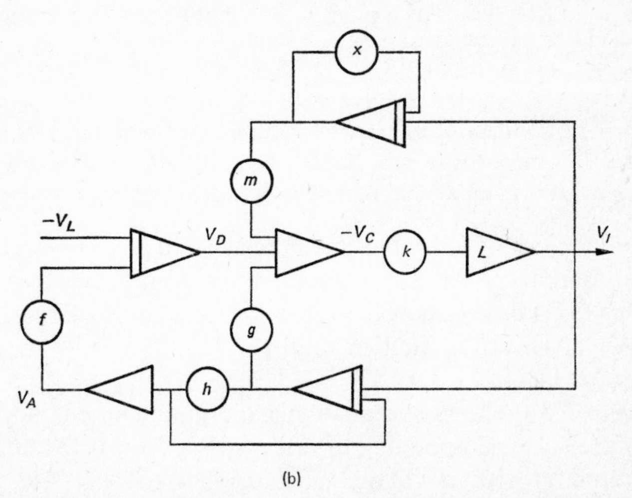

Figure 62 (a) Block diagram, and (b) analogue computer models of drinking. (*Source: McFarland*, 1971)

Water content in the gut is the integral of the net flow of water into the gut. Thus the bottom transfer function $1/s$ relates gut content to flow into the gut. Net flow into the gut is the difference between the rate of water intake I and the rate of absorption from the gut A. The bottom summing junction forms this difference. Absorption rate A is linearly related to gut content by the constant h.

The top left of the diagram shows that if the net rate of water loss is integrated, the debt is obtained, in other words, the error in the body fluids. The $1/s$ term performs this operation. Net rate of water loss will be rate of removal minus rate of gain from the gut, and a summing junction at the top left subtracts rate of gain from rate of loss L. The gain f represents the 'hydrating power' of the water that has been absorbed from the gut. For distilled water the gain f would be a maximum, whereas it would be less for a saline solution. The signal marked 'command' represents the thirst stimulus, and as may be seen is dependent not only upon the debt, but is a function of three variables. The gut content exerts an inhibitory effect via the gain g. The gain K relates the command to the rate of drinking I.

The model includes oral feedback. The transfer function $m/(s+x)$ provides a decaying memory of the amount of water drunk, and this serves as a positive-feedback loop. According to McFarland (1971):

> . . . oral factors provide positive feedback, which serves to maintain drinking in the face of competing stimuli from other motivational systems. In other words, oral factors serve to reinforce the ongoing behaviour and maintain its momentum until satiation is complete.

The conclusions of McFarland in relation to positive feedback in the Barbary dove are somewhat different from those of Toates (1971) based upon the rat, a subject that is discussed in Chapter 8. However, fundamental species differencies may be important.

Figure 62(b) shows the analogue simulation of the system. A voltage V_L, corresponding to rate of water loss, is fed into an integrator to give a voltage proportional to debt V_D. The voltage $-V_c$ is the algebraic sum of positive voltages representing the debt and oral feedback, and the negative voltage representing gut inhibition. The voltage V_I that represents rate of ingestion is fed back as an input to the bottom integrator, the output of which is gut water. A potentiometer having a setting h provides a linear relationship between gut content and absorption rate V_A. The term f is a potentiometer setting corresponding to the hydrating power of the absorbed water, while g is another potentiometer setting that represents the inhibitory effect of gut water on the thirst mechanism. A potentiometer setting x determines the time constant of the decay of the memory of drinking, while another potentiometer of gain m determines the excitatory effect on the thirst mechanism.

McFarland also introduced a non-linear feature into the simulation. Ingestion rate V_I was only allowed to reach a certain maximum value, and this was done by preventing the output of the amplifier L from rising above a particular voltage. Such a non-linearity has just been described, and in the present case it means that rate of intake is proportional to debt so long as debt is below the saturation point, but for debts larger than this there is a maximum rate of ingestion. This forms the analogue of a real experiment in which birds work for water in a Skinner box, the maximum rate at which they may obtain it being limited by experimenter. A comparison between the bird's performance and the computer prediction is shown in Figure 63(a). It may be seen that as the maximum rate of water intake is reduced, then the bird's actual rate stays at this maximum for a longer time. In the case of 0 s time-out the rate slows up relatively quickly.

By reference to behavioural results, McFarland was able to estimate the value of the gain f for a saline solution. For distilled water, f is assumed to have the value of 1·0, since presumably a gain of 1 cm^3 of water balances a loss of 1 cm^3. The average consumption of water is 11·0 cm^3 for 48 h deprived birds, whereas if following the same deprivation period 0·5 per cent saline is given, an average of 17·5 cm^3 is consumed. The hydrating power of saline is therefore calculated to be 11/17·5 or 0·63. If this value is substituted into the computer simulation, a close fit between the experimental result and simulated result occurs, as Figure 63(b) demonstrates.

Returning to the construction of the analogue simulation, it should be noted that there is a close correspondence between the analogue computer terms and the block diagram. One additional unit is shown in the analogue simulation apart from the non-linearity L and this is the amplifier at the bottom left. An additional unit such as this needs to be employed to obtain the necessary sign changes.

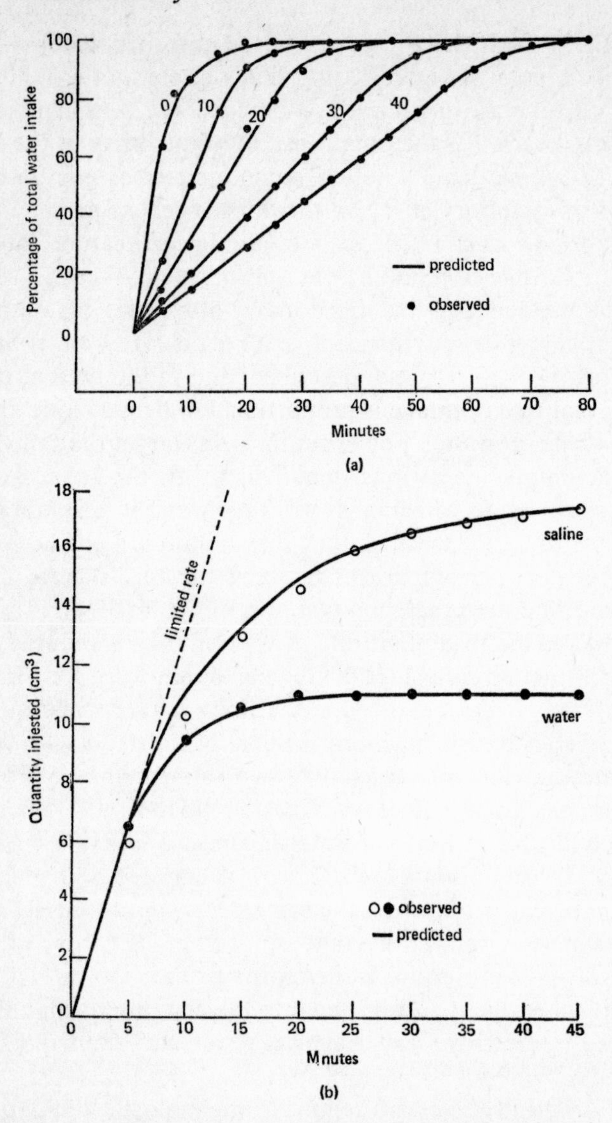

Figure 63 Comparisons between an analogue computer model and the experimental performance of birds. (a) The rate of water availability has been limited following a period of deprivation. The times indicate the time in seconds during which reward could not be obtained following the previous reward. (b) The consumption of water and saline following 48 h *of water deprivation. The model is based on the fact that saline has a hydrating power of* 0·63. (*Source: McFarland and McFarland,* 1968)

7 FREQUENCY RESPONSE METHODS

It was explained in Chapter 4 that information about the nature of a system may be gained by applying step and ramp disturbances to it, and then observing responses to these stimuli. In analysing a system, one usually needs to extract as much information as possible, and a technique that is complementary to step and ramp testing is the method of frequency response analysis.

The idea of frequency response ought not to be unfamiliar to the biologist. To give just one example, a sensitivity curve of the ear shows auditory sensitivity as a function of the frequency of the sound stimulus. Sensitivity rises as frequency increases until a maximum is reached, whereupon the sensitivity decreases. It reaches zero at about 20 000 hertz, the exact value depending upon the age of the person. Some animals exhibit a higher frequency range than man. When discussing frequency response we are then building upon concepts that are, up to a point, established in biology.

A frequency response test involves applying a sinusoidally oscillating test stimulus to the system and measuring the resultant response. A sinewave is shown in Figure 64, and is described by the equation

$$K \sin \omega t$$

where K = gain constant, ω = radian frequency, and t = time. An explanation of how a sinewave is generated was given in Chapter 2.

If the system is linear, the output will oscillate sinusoidally at the same frequency as the input stimulus when we apply a sinewave to the input. However, in general the output sinewave will be changed in phase with respect to the input sinewave. The

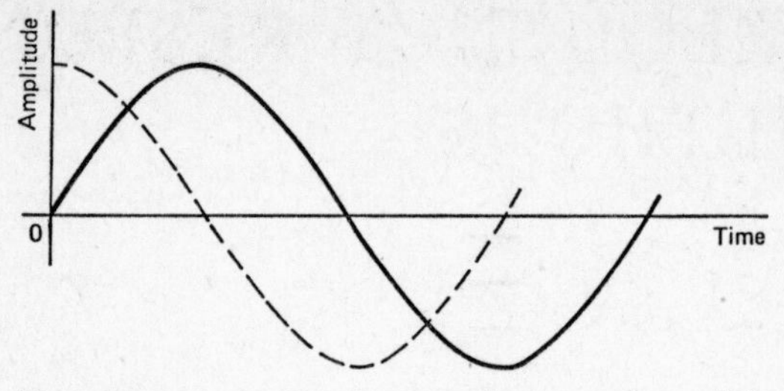

Figure 64 Two sinewaves having the same frequency but a difference in phase (for explanation see text).

meaning of a phase shift is illustrated in Figure 64. Although the two sinewaves are oscillating at the same frequency, they are not doing the same thing at the same point in time. In this example when one is at either its maximum or its minimum the other is at zero. They may be superimposed simply by moving one of them through a quarter of a cycle, or in other words 90°, and therefore we would say that these two sinewaves are 90° out of phase.

Let us suppose that the continuous line represents the input which we have applied to the system, while the broken line is the output. At the particular frequency shown, Figure 64 indicates that the system has the characteristic of giving a 90° phase advance to the signal, but leaving its amplitude unchanged. The output sinewave has the same amplitude as the input sinewave. The shift is one of phase advance since at any given time the output is doing what the input will be doing at a quarter of a cycle later.

Suppose we were to double the input frequency, it might then be found that the phase difference would increase to, say, 120° and the amplitude of the output sinewave increase so that it is now 1·5 times the size of the input sinewave. At this new frequency we would say that the characteristic of the system is to give a phase advance of 120° and a gain of 1·5. Gain is the ratio of the height of the output sinewave to that of the input sinewave. To carry out a frequency response analysis we examine this ratio at a particular frequency and find the gain of the system. We also examine the phase shift at the same frequency. We then change the frequency and repeat the observations. The variation in gain and phase shift with changes in frequency depends upon the nature of the system relating input to output, and hence this information is used to reveal the system's identity.

In Chapter 2 the concept of a vector was introduced, and it was explained that a vector is a quantity having both direction and magnitude. Vector notation is one means employed to represent the frequency response of a system, since the response may be completely characterised in terms of magnitude, i.e. the gain of the system and direction or phase shift. This is most easily explained by reference to Figure 65, which shows the frequency response that we have just discussed.

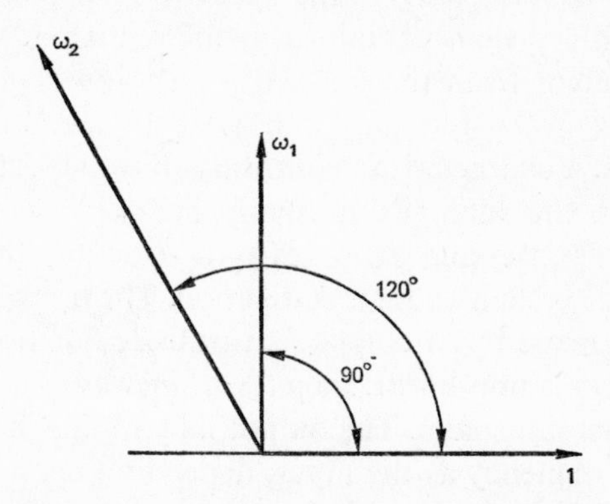

Figure 65 *At frequency* ω_1 *the phase-shift is* 90° *and the gain* 1·0. *As the frequency is changed to* ω_2 *the gain increases to* 1·5 *and the phase-shift to* 120°.

At the first frequency (we will call it ω_1) the gain was 1 and the phase shift 90° leading. This is represented by a line the length of which is proportional to 1 and which makes an angle of 90° to the horizontal. The symbol ω_1 is written alongside the vector to indicate that it was obtained at frequency ω_1. Phase advance is measured in an anti-clockwise direction, and all angular measurements are made with respect to the positive x-axis. Sometimes, as in Figure 65, a vector of length 1 is drawn on the x-axis to establish a reference.

When the input frequency was increased to ω_2 it was found that the phase advance increased to 120° and the gain became 1·5. Therefore, corresponding to frequency ω_2, a vector is drawn having a length of 1·5 and a phase angle of 120°.

In carrying out a frequency response analysis of a system, a series of signals at various frequencies is applied and at each frequency a vector drawn to show gain and phase. If the tips of the vectors are joined together we obtain a frequency portrait of the system, known as a frequency response diagram. In practice the vectors themselves are omitted and only the locus through their tips is drawn.

When one knows the rules for interpreting the frequency response diagram, and particularly when it is viewed in conjunction with the response to step and ramp stimuli, it can often reveal the component parts of the system very convincingly. For instance, if the system contained a differentiator, this should become apparent from the frequency response diagram. Each component transfer function that goes to form the overall system makes a characteristic contribution to the diagram. The problem that the biologist normally has is to examine the frequency response and from this to estimate the transfer function of the system under investigation. The transfer function will often be made up of component transfer functions.

If the system is non-linear, then for a sinewave input the output will not be sinusoidal. The output may or may not oscillate at the same frequency as the input, depending upon the nature of the system. For instance, if we were to apply a sinewave to a saturation non-linearity of the kind shown in Figure 38 on page 102, the output would oscillate at the same frequency as the input. The amount by which it would depart from a sinewave would depend upon the size of the input. If small signals were used then we would be operating over the linear range and the output would be sinusoidal. However, if the input were to invade the saturation region, the output could not be sinusoidal since the maximum point of the sinewave input would not be transmitted. In many cases, especially with biological control systems, linear frequency response methods are employed as a first appoximation. However, the warning given in Chapter 4 that the use of linear techniques for systems having severe non-linearities can be positively deceptive, must be repeated here. There are rigorous and highly sophisticated mathematical techniques available for non-linear frequency response analysis, but they are beyond the scope of this book.

Open- and Closed-Loop Tests

The problem of cutting or opening a feedback loop was discussed in Chapter 5. We are now in a position to appreciate the mathematics of opening a feedback loop.

Consider the negative-feedback control system shown in Figure 66(a). For a sinewave, as for other input stimuli, the

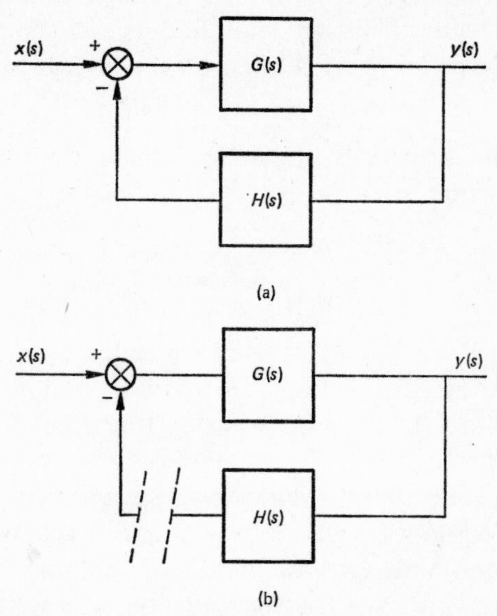

Figure 66 *The feedback loop of the system shown in* (*a*) *has been cut in* (*b*).

relationship between input and output is given by

$$\bar{y}(s) = \frac{G(s)}{1 + [G \times H(s)]} \times \bar{x}(s)$$

where $\bar{x}(s)$ = input, $\bar{y}(s)$ = output, $G(s)$ = forward loop transfer function, and $H(s)$ = feedback loop transfer function.

This equation was developed in Chapter 6. For the purpose of identifying the components of the system, the reader will appreciate that the relationship between input and output is not particularly simple. That is to say $G(s)/[1+G \times H(s)]$ reveals neither $G(s)$ nor $H(s)$ very readily. In some situations it is precisely $G(s)$ and $H(s)$ that we wish to know.

Let us suppose that we were able to cut the feedback loop so as to prevent the output from being fed back to the input, as shown in Figure 66(b). If we now apply a sinewave to the input, and measure the output response, the relationship between input and output is defined simply by $G(s)$. Even if we were to allow the signal to travel round the loop, and measure the response at the point at which the loop is cut, the transfer function between input and the response is $G \times H(s)$, which is considerably simpler than $G(s)/[1+G \times H(s)]$. For this reason, as well as others which will be discussed later, frequency response measurements are very often carried out with the loop broken.

Predicting the Frequency Response from a Knowledge of the Transfer Function

If we want to be able to interpret experimentally obtained frequency response plots it is necessary to know what the frequency response of various transfer functions looks like. In other words, if we can construct frequency response diagrams from known transfer functions, we are in a good position to do the reverse, i.e. to estimate transfer functions from experimentally obtained frequency response plots.

Let us imagine that we have a transfer function given by $G(s) = s$. We know from Chapters 3 and 4 that this transfer function corresponds to a differentiator. Suppose we were to apply a sinewave to this transfer function, and we will assume that this is the sinewave shown by the continuous line in Figure 64. At time zero the sinewave is changing from a value of zero to a positive value, and so its time derivative at this point is some positive quantity. As from time zero its rate of change slows up, until at the point when the sinewave is at its maximum it has a rate change of zero. After the maximum, the rate of change becomes negative. In fact if we estimate the derivative of the continuous curve at various points in time, we will find that the derivative follows exactly the form of the function shown dotted in the same diagram, i.e. a cosinewave. A cosinewave is 90° ahead of a sinewave, and that is the only difference. The effect of differentiating is then to introduce a 90° phase advance. It will be recalled that information on phase and gain may be represented by a vector diagram. In this way we could represent the effect of the transfer function $G(s) = s$ by a vector lying along the positive imaginary axis, since it rotates the vector by 90°. In terms of complex numbers, discussed in Chapter 2, the effect of the transfer function $G(s) = s$ is to multiply the vector by j.

We must now consider what happens to the amplitude as a result of differentiation, i.e. what is the length of the vector which we have located on the j-axis? The proof will not be developed here, but the derivative of $\sin \omega t$, where ω is the radian frequency, is given by $\omega \cos \omega t$. Thus the effect is to multiply the signal by ω as well as to shift it 90° forward. This is what we would expect. A signal that is oscillating at 20 Hz obviously has a higher rate of change at any point in the cycle than one exhibiting a 0·2 Hz oscillation. Thus to differentiate means to move the phase forward by an angle of 90° (to multiply it by j) and to multiply the amplitude by ω.

In other words we could have replaced the s by the term $j\omega$ and we would have characterised the system's performance. As far as frequency response is concerned – and only that far – the transfer function of the differentiator $G(s) = s$ may be replaced by $j\omega$, where ω is the radian frequency of the sinewave that we apply. As a general rule for any transfer function, we may substitute $j\omega$ for s when we employ sinusoidal inputs. If the reader is interested in a mathematical justification of this, he or she should consult a text on the mathematics of control theory. The example given was a demonstration and not a proof.

Having replaced s by $j\omega$ throughout the transfer function, we then plot the transfer function as a function of ω. To do this we select values of ω and fix several points, then a locus is drawn running through the points. The differentiator as a transfer function is

$$G(s) = s$$

therefore $$G(j\omega) = j\omega$$

This is a purely imaginary number. Therefore, for all values of ω, the frequency response or, as it is often called, the Nyquist plot, will lie along the positive imaginary axis. This is shown in Figure 67(a). We select some values of ω and fix the locus as far as magnitude is concerned. When ω is zero the value of $j\omega$ is also zero, and when ω is infinity $j\omega$ has a magnitude of infinity.

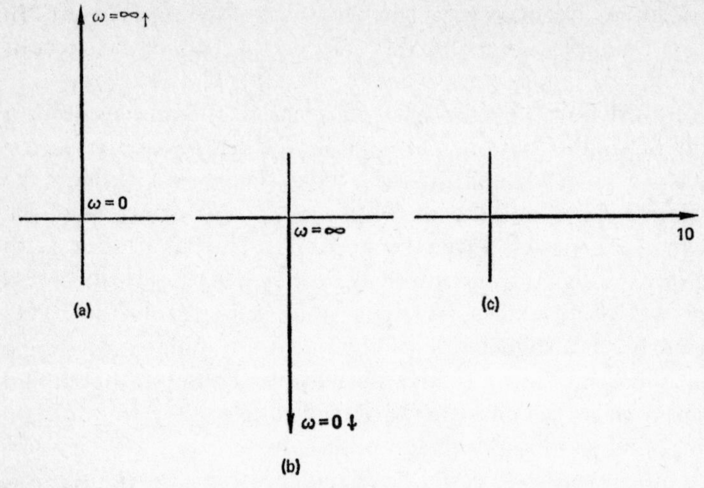

Figure 67 The frequency response of some transfer functions. (a) The differentiator $G(s) = s$, *(b) the integrator* $G(s) = 1/s$, *(c) the pure gain* 10.

Thus the locus goes from zero to infinity along the positive imaginary axis.

A few examples should serve to familiarise the reader with the procedure.

Example 18

What does the Nyquist plot for an integrator look like? From Chapter 4 we know that the transfer function of an integrator is given by

$$G(s) = 1/s$$

therefore $G(j\omega) = 1/(j\omega)$

It is best to get rid of terms in j from the denominator and we can do this by multiplying top and bottom by j.

$$G(j\omega) = \frac{j}{j^2\omega}$$

where $j^2 = -1$, therefore

$$G(j\omega) = -\frac{j}{\omega}$$

As with the differentiator, this is an imaginary number, but in the present case it will always be negative. When ω is zero $G(j\omega)$ has a magnitude of infinity, and when ω is infinity $G(j\omega)$ is zero. Thus its behaviour is precisely the opposite to that of the differentiator. Its Nyquist plot is drawn in Figure 67(b), which shows that an integrator gives a 90° phase lag to a sinewave.

Example 19

What is the Nyquist plot for a pure gain of 10? For all values of ω the transfer function is simply given by 10, a real number. Therefore a single vector of length 10, lying along the positive real axis, as shown in Figure 67(c), characterises the performance of the transfer function at all frequencies. No phase shift is ever introduced, the output always being 10 times the input.

Example 20

Sketch the Nyquist plot corresponding to the transfer function

$$G(s) = \frac{sT}{1+sT}$$

where T is a constant

$$G(j\omega) = \frac{j\omega T}{1+j\omega T}$$

It is generally easiest to begin by putting the values $\omega = 0$ and $\omega = \infty$ into the equation, and determining these two extremes of the range. For large values of ω, the 1 in the denominator becomes insignificant by comparison with the $j\omega T$. A complex number of imaginary part ∞ and real part 1 is almost the same as a pure imaginary number. At ω equals ∞ we have the purely real number 1, and this is drawn on Figure 68.

When ω is very small the term $j\omega T$ in the denominator becomes insignificant with respect to the 1. This means that the denominator is approximately 1, and the transfer function approaches the value $j\omega T$ for low frequencies. As ω approaches zero the transfer function approximates to a purely positive imaginary number of

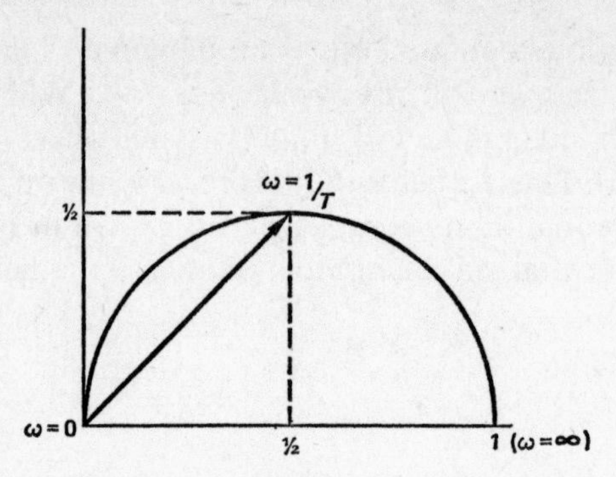

Figure 68 *Frequency response of the transfer function* $G(s) = sT/(1+sT)$.

length zero. Thus the locus emerges from the origin along the positive imaginary axis.

It is necessary to examine the response for at least one intermediate frequency in order to sketch the Nyquist plot. To do this it will be easier if we eliminate imaginary terms from the denominator, by multiplying top and bottom by $1-j\omega T$.

$$F(j\omega) = \frac{j\omega T\,(1-j\omega T)}{(1+j\omega T)(1-j\omega T)}$$

$$= \frac{j\omega T - j^2\omega^2 T^2}{1-j^2\omega^2T^2}$$

since $j^2 = -1$

$$= \frac{j\omega T + \omega^2 T^2}{1+\omega^2 T^2}$$

Life is made relatively simple for us if we substitute the value $\omega = 1/T$, since $\omega T = 1$.

$$F(j\omega) = \frac{j(1/T)T + (1/T^2)T^2}{1+(1/T^2)T^2}$$

$$= \frac{j+1}{1+1}$$

$$= 1/2 + j/2$$

At a frequency $\omega = 1/T$ the vector is $\frac{1}{2}+\frac{1}{2}j$ and this now enables us to estimate the Nyquist plot.

The Bode Plot

Although the Nyquist diagram serves a very useful purpose, under some circumstances it has its disadvantages. For instance, it is not easy to represent on the diagram the frequency at which a particular point of the locus was obtained. Secondly, given the Nyquist plot of two or more individual transfer functions it is difficult to obtain readily the frequency response of the combination. Both of these problems may be overcome by the use of the Bode diagram. Also, a rough Bode plot may be sketched very quickly from a knowledge of the transfer function.

The vectors of which the Nyquist plot is composed contain two pieces of information, i.e. the phase shift and the gain. Instead of incorporating this information into a single diagram, for the Bode plot we construct two diagrams, one showing gain and the other phase, each as a function of frequency. Frequency is plotted logarithmically as shown in Figure 69. In this way the high-frequency range is compressed and the low-frequency range expanded, which is generally more useful than a linear scale. Gain is not plotted directly as such, but rather the gain in decibels is plotted. This is defined as

$$\text{gain (decibels)} = 20 \log_{10} (\text{gain})$$

The constant term 20 is not arbitrary, but arises in the adaptation of this method from radio engineering. Phase is plotted directly in degrees.

In Figure 69(a) we have the Bode plot for a transfer function G_1, and in (b) for a second transfer function G_2. Suppose that we needed the frequency response of the transfer $G_1 \times G_2$. Since gain in db is logarithmic and the logarithm of a product is the sum of the logarithms of its components (see Chapter 2), we simply add the two individual plots together. For phase, the total phase shift is the sum of the component phase shifts, and we can just add without using logarithms. Figure 69(c) shows the frequency response of $G_1 \times G_2$ which is simply given by $G_1 + G_2$, both for gain and phase.

A few examples are essential in order to understand the Bode diagram.

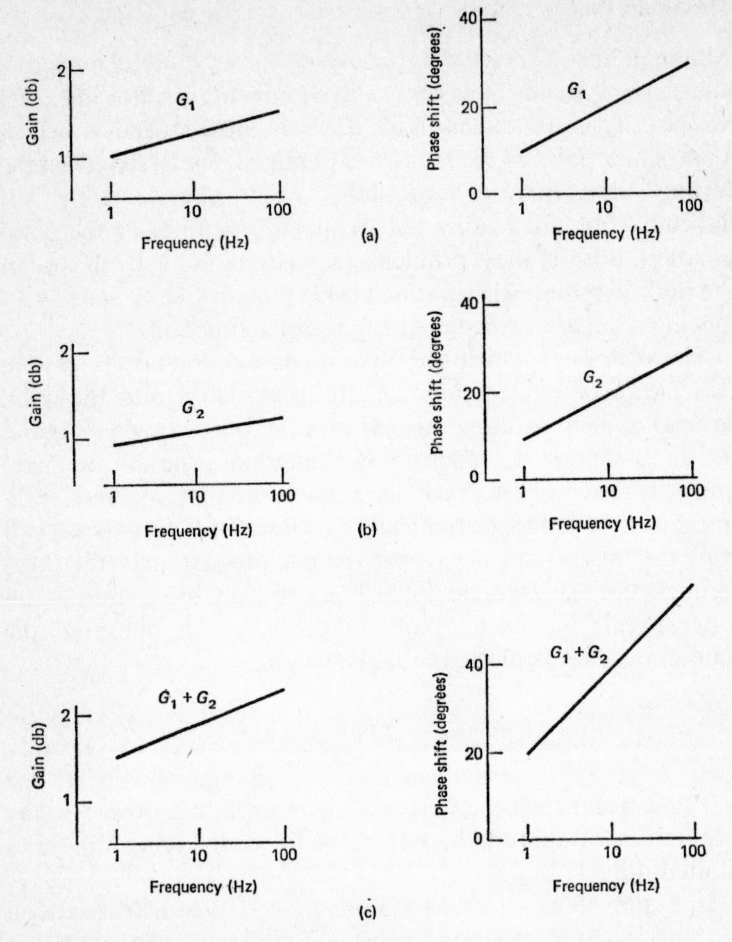

Figure 69 *If we add the Bode plots shown in (a) and (b) we obtain in (c) the Bode plot of the combination G_1+G_2. This represents the frequency response of G_1 and G_2 when connected in series.*

Example 21

Sketch the Bode plot of a pure gain K.

Figure 70 shows the plot. The gain is K and so the gain in decibels is given at all frequencies by $20 \log_{10} K$, which is itself a constant. Since the transfer function is a purely real number the phase shift is zero for all frequencies.

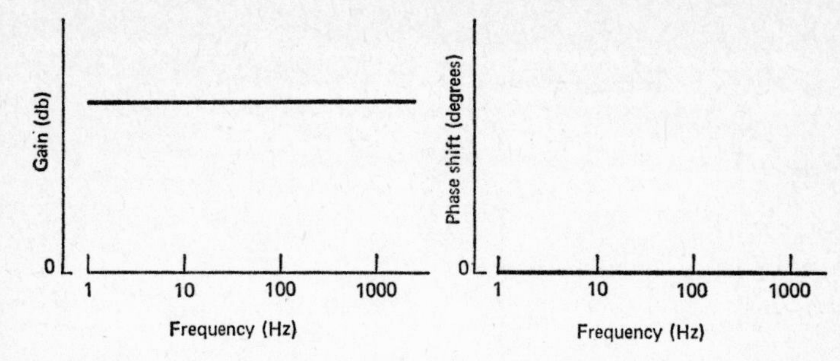

Figure 70 Bode plot for a pure gain K, where K is greater than 1.

From the way in which Figure 70 is drawn we can immediately say something about K. What is it? The answer is that K must be greater than 1, otherwise $20 \log_{10} K$ would be negative. When K is 1, $20 \log_{10} K$ equals zero. Thus multiplying a signal by a transfer function of 1 leaves it unchanged, just as adding a plot of zero to the Bode diagram does not alter anything.

Example 22

Sketch the Bode plot for the transfer function $G(s) = 1/(1+sT)$.

This is the transfer function of an exponential lag element, and is a very important component in many biological systems.

$$G(s) = 1/(1+sT)$$

$$G(j\omega) = 1/(1+j\omega T)$$

$$= (1+j\omega T)^{-1}$$

$$\text{gain (db)} = 20 \log_{10} (\text{gain})$$

In vector notation, gain is the length of the vector, of which the real and imaginary terms form components. The magnitude of the vector is, as shown below, indicated by two vertical lines enclosing the complex number.

$$\text{gain (db)} = 20 \log_{10} |(1+j\omega T)^{-1}|$$

We can bring the -1 to the front.

$$\text{gain (db)} = -20 \log_{10} |(1+j\omega T)|$$

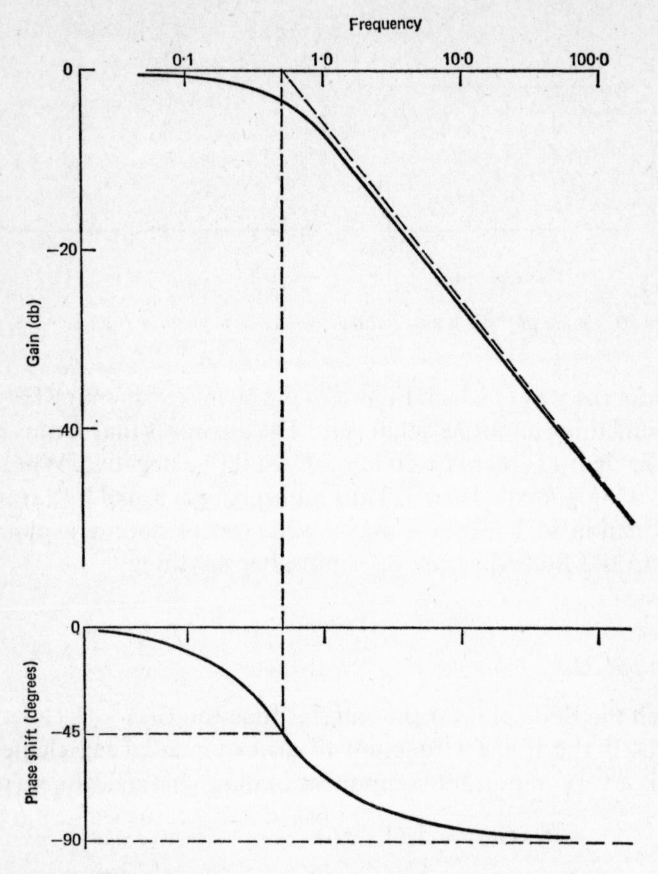

Figure 71 *Bode plot of the transfer function* $G(s) = 1/(1+sT)$.

The magnitude of the vector is given by Pythagoras' theorem, where the sides of the right-angled triangle are the real and imaginary components of the complex number. Thus

$$\text{gain (db)} = -20 \log_{10} (1^2+\omega^2T^2)^{\frac{1}{2}}$$
$$= -20 \log_{10} (1+\omega^2T^2)^{\frac{1}{2}}$$

For very low values of ω we can ignore the ω^2T^2 with respect to the 1, and this means

$$\text{gain (db)} \simeq -20 \log_{10} 1 = 0$$

For very large values of ω we can disregard the 1 and thus

$$\begin{aligned}\text{gain (db)} &\simeq -20 \log_{10} (\omega^2 T^2)^{\frac{1}{2}} \\ &= -20 \log_{10} \omega T \\ &= -(20 \log_{10} \omega + 20 \log_{10} T) \\ &= -(20 \log_{10} \omega - 20 \log_{10} 1/T)\end{aligned}$$

The term $20 \log_{10} (1/T)$ is a constant. For the moment let us just consider the term $-20 \log_{10} \omega$. As the frequency ω is increased by a factor of 10 (i.e. a decade) then $\log_{10} \omega$ will increase by 1, and hence $-20 \log_{10} \omega$ will decrease by 20. The slope is -20 db/decade, which fixes the rate at which the decibels drops off, but does not locate the plot. Sometimes frequency is plotted in terms of octaves, i.e. $\frac{1}{2}$, 1, 2, 4, at equal intervals rather than 1, 10, 100. When octaves are used, as the frequency doubles, $\log_{10} \omega$ increases by 0·3. Similarly, $-20 \log_{10} \omega$ decreases by 6, and so the slope is -6 db/octave. Thus the term $-20 \log_{10} \omega$ may be said to have a gain of -20 db/decade or -6 db/octave. They both amount to the same thing. A gain of 1 is 0 db, and $\frac{1}{2}$ is -6 db, while a gain of 2 is 6 db.

In order to locate the -20 db/decade slope on the Bode diagram we must consider the term $(20 \log_{10} \omega - 20 \log_{10} 1/T)$. When ω equals $1/T$ this term has the values zero; this fixes a point through which the slope passes. As Figure 71 shows, we can draw two asymptotes to which the plot tends at the extremes of the frequency range. At low frequencies it tends to zero and at high frequencies to the ωT asymptote. When the frequency approaches either of these extremes, the simplifying assumptions provide greater accuracy. However we must fix at least one point at an intermediate frequency, for which neither of our simplifying assumptions may be applied. If we choose the frequency $\omega = 1/T$, we will find that the mathematics is made easy for us.

$$\begin{aligned}\text{gain (db)} &= -20 \log_{10} (1 + \omega^2 T^2)^{\frac{1}{2}} \\ &= -20 \log_{10} [1 + (1/T^2) T^2]^{\frac{1}{2}} \\ &= -20 \log_{10} 2^{\frac{1}{2}} \\ &= -3 \text{ db}\end{aligned}$$

At $\omega = 1/T$ gain has the value of -3 db, and thus we can construct the gain curve from this point and the asymptotes at the extremes of the range.

This now leaves the phase curve to be drawn, and again it is easiest to consider the extremes of the frequency range. When the frequency is very low the expression $1/(1+j\omega T)$ approaches $1/1$, which is a purely real number. Therefore, at low frequencies the phase shift is zero.

At very high frequencies we can ignore the 1 with respect to the $j\omega T$, and so we have $1/j\omega T$, which becomes $-j/\omega T$ after multiplying top and bottom by j. In this way we obtain a purely imaginary number, but with a negative sign, which means a 90° phase lag at high frequencies.

In order to examine an intermediate frequency we first eliminate j from the denominator.

$$
\begin{aligned}
G(j\omega) &= \frac{1}{1+j\omega T} \\
&= \frac{(1-j\omega T)}{(1+j\omega T)(1-j\omega T)} \\
&= \frac{1-j\omega T}{1+\omega^2 T^2} \\
&= \frac{1}{1+\omega^2 T^2} - \frac{j\omega T}{1+\omega^2 T^2}
\end{aligned}
$$

When real and imaginary parts are equal this means that the phase lag is 45° (consider the vector in terms of a right-angled triangle having equal sides). This will occur when $\omega T = 1$, as we may readily see by substituting in the above equation. For ωT to equal 1 then ω must equal $1/T$. This now gives us three points, and from them we can construct the phase curve.

Stability

Consider the possible responses of a system to a step change in input which are shown in Fig. 72. Part (a) represents a sluggish response, and we would say that a system exhibiting such behaviour is over-damped. Part (b) shows a critically damped system in which the response is as fast as is possible before

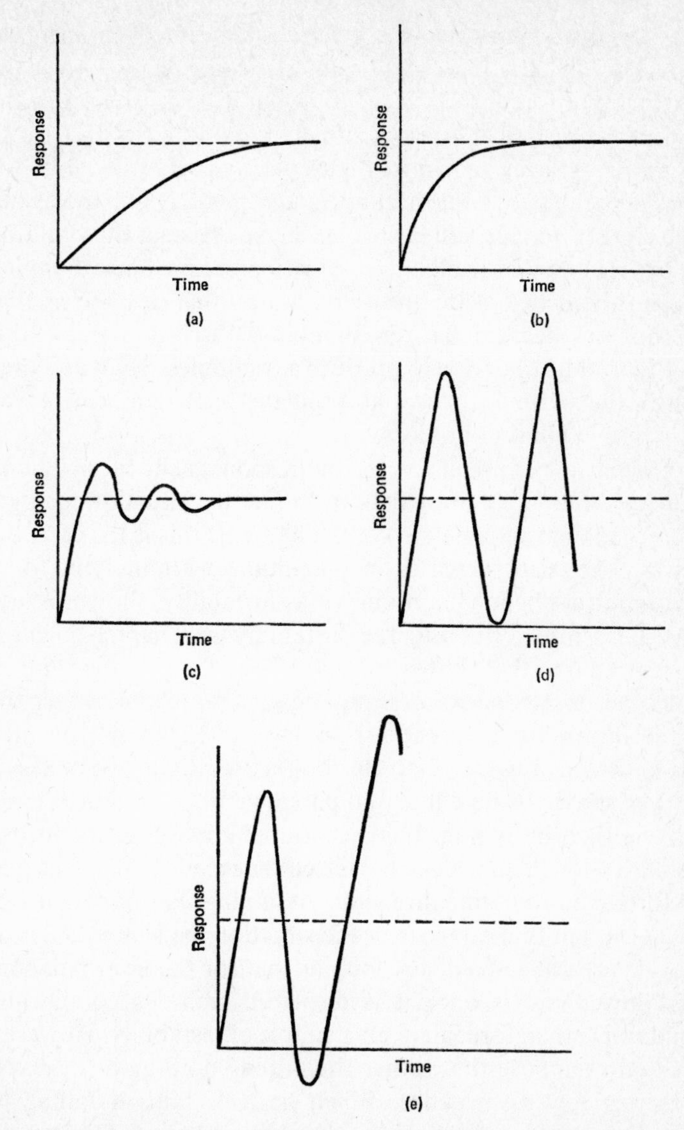

Figure 72 The possible responses of a system. (a) Over-damped, (b) critically damped, (c) the response exhibits an overshoot, (d) damping is zero, (e) instability.

overshoot occurs. Overshoot is shown in the response of (c) where the system only comes to rest after an oscillation. This arises because the output has a certain amount of momentum that carries it past the equilibrium position. After a time the counteracting force is sufficient to swing it back, but again the momentum carried it past its target. Finally it comes to rest. The swinging lever shown in Figure 44 (page 119) exhibits the same performance if the friction in the pivot is not too great. If the system is made still less sluggish, we arrive at the situation shown in (d) where a prolonged oscillation occurs. Since damping refers to the decay of the transient, and in this case the oscillation shows no decay, it has zero damping. We could simulate this if in Figure 44(a) we were to produce a frictionless pivot and then disturb the system. It would oscillate back and forth, the oscillation exhibiting no decay.

A system may exhibit any of the responses shown in Figure 72, depending on its construction. In fact the situation may get even worse than shown in (d) of the figure. If this is the case, the output oscillation increases in magnitude without limit for a finite input, a phenomenon known as instability, shown in (e).

We have already considered instability in Chapter 5, but a different kind to that shown in Figure 72(e). To give an example, the learner tightrope walker may be seen to exhibit either the kind of instability described for the lever of Figure 44(b) or the kind shown in Figure 72(e). So long as the tightrope walker's centre of gravity is on a line that passes vertically upwards from the rope then he is safe. If his centre of gravity moves to one side of this line then a force is exerted that tends to displace him still further in the same direction. Assuming that our tightrope walker does nothing to correct the situation, he is in exactly the equivalent positive-feedback loop as that for the lever balanced at its lower end is once it is displaced from its equilibrium position. He therefore falls – an example of instability. However, let us now suppose that rather than doing nothing he attempts to correct any disturbance. When he feels himself falling he takes an action to return himself to the vertical. Due to the inevitable time delays in the system, the corrective action is still being taken when he is back in the equilibrium position and so he is carried past, thereby creating an error of the opposite sign. It is possible that the momentum is sufficient to create an even larger error than the original one. This then calls for another reaction to restore his equilibrium; this creates an even bigger error. After a series of such oscillations, he falls. Presumably the master tightrope walker has learned the correct responses to be associated with any particular error in a way comparable to the skilled motorist who was described in Chapter 5.

The responses shown in Figure 72(a–c) are all stable. The response of (d) is on the border of instability, since if we were to decrease the damping, i.e. to make it negative, the oscillation would increase in amplitude and then be unstable.

When designing a control system the engineer wishes to know what kind of transient response to expect from it. Obviously no one would, except by accident, build a system which is inherently unstable. Although, of course, the task is quite different in biology; the question of stability and instability is central to many problems.

We can predict the stability of a system by inspecting its open-loop frequency response. This is usually the reason for opening the feedback loop in engineering systems, since the stability of the system on closed-loop may be predicted from the safe position of carrying out an open-loop test, as instability is a phenomenon associated with closing the loop. However, occasionally unstable open-loop systems may be stabilised by feedback.

Let us suppose that there is a certain frequency at which the phase shift around the loop is $-180°$. This means that the signal fed back to the summing junction is 180° out of phase with the input. The summing junction causes the signal to move another 180° out of phase (because of the minus sign), which means that the input and the feedback signal are now exactly in phase. If the gain around the loop is greater than 1 this means that any disturbance within the system at the frequency for which the open-loop phase shift is 180° will tend to be self-generating, and a prolonged oscillation will result. It is not necessary to apply the particular frequency at the input, due to random noise in systems a disturbance at the critical frequency will arise. If the loop gain is greater than 1 this means that with each passage around the loop the signal is amplified, and so the oscillation will increase in strength on each cycle. Instability will result. Thus we look at the open-loop frequency response, and if there is a phase shift of 180° at any frequency, the gain at that point must be less than 1 for stability. This is shown on the Nyquist diagram of Figure 73.

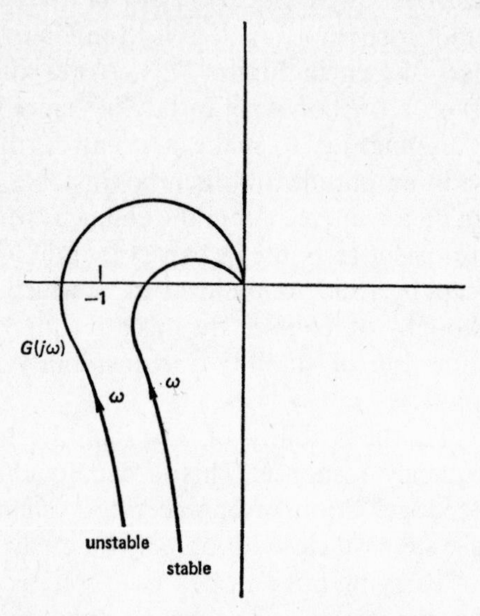

Figure 73 Nyquist plot.

The reader must be warned that the detailed mathematics of stability is far more involved than might have been suggested here, and sometimes this simple test for stability will prove to be inadequate.

Frequency Response of Biological Systems

As the basis of the method of frequency response analysis has been presented, we are in a position to consider the application of these techniques to biological systems. Three examples will be given of where frequency response methods have been employed in biology. Space permits only a summary of the experiments, and the student is strongly advised to read the original article in each case. This is particularly true of the pupil light reflex study of Stark and Sherman, which forms a landmark in biological systems analysis.

THE PUPIL LIGHT REFLEX OF THE HUMAN EYE

The amount of light reaching the retina of the eye is determined by the size of the pupil aperture. When the light intensity is higher than is desired, the pupil light reflex causes a contraction of the pupil in order to reduce the amount of admitted radiation. Conversely, in dim illumination the aperture enlarges so that more light falls on the retina. The pupil light reflex was investigated as a servo-mechanism by Stark and Sherman (1957).

Figure 74 shows a simplified representation of the system. The

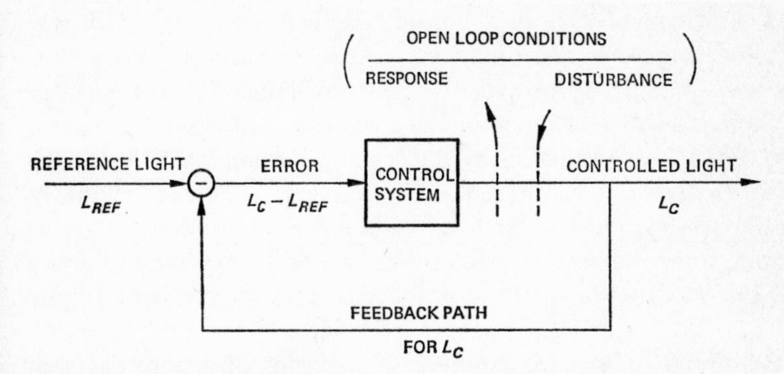

Figure 74 The pupil light reflex. (*Source: Stark and Sherman*, 1957)

controlled quantity is the amount of light falling on the retina (L_c). There is also a reference light flux quantity (L_{REF}), L_c being compared with L_{REF}. The difference, or error $L_c - L_{REF}$ is the signal which actuates the control system.

Stark and Sherman measured pupil area by reflecting infra-red light from the iris to a photocell. The pupil light reflex is not disturbed by doing this since the eye is insensitive to infra-red radiation. Thus both the light stimulus and the infra-red source may be directed into the eye simultaneously. The system was calibrated in terms of the deflection on an oscilloscope as a function of pupil area. For a large pupil, relatively little infra-red radiation is reflected onto the photocell, while a small pupil means a large reflecting surface and hence a large deflection on the oscilloscope.

Stark and Sherman designed a technique for opening the feedback loop which did not involve any interference with the system. This was done by focusing the light stimulus so that the light entering the eye had a diameter smaller than the smallest diameter of the pupil. Thus any changes the pupil made were unable to influence the amount of light falling on the retina.

The disturbance was introduced at the output end of the system by modulating the intensity of the light falling on the retina sinusoidally, and the response, in the form of pupillary changes, was measured. Small signals were employed so that linear techniques could be used for the analysis.

Figure 75(a) shows the open-loop frequency response plotted in the form of a Bode diagram. The low-frequency gain was found to be 0·16. At high frequencies the gain falls off with a slope of −18 db/octave. As was explained earlier in this chapter, a single exponential lag is responsible for −6 db/octave, and so it was estimated by Stark and Sherman that three such lag elements were present, each having a time constant of approximately 0·1 s. Each of these elements produces a maximum phase lag of 90°, and so three would be expected to give a lag no greater than 270°. In practice, as may be seen from Figure 75(a), at 4 hertz the phase lag is 540°. Stark and Sherman accounted for the remaining 270° in terms of a pure delay of 0·18 s. (See Chapter 4 for explanation of the pure delay.) This value was obtained experimentally in response to a step input. The pure delay has absolutely no effect as far as gain is concerned since it merely delays the signal, but it gives an ever increasing phase lag as the frequency is increased. Since a pure delay does not change the gain, its Nyquist plot is a circle of radius 1, traced out by a vector of length 1 which rotates about the origin. It is interesting to note that the magnitude of the pure delay obtained in response to an unpredictable step input is equally applicable to the behaviour of the system to a predictable sinusoidal input. In controllers that exhibit learning (see, for instance, the accommodation control study of Stark, Takahashi and Zames (1965), described in Chapter 5) the delay in response to a step input stimulus is too long to be applied to a predictable stimulus. This demonstrates that the pupil light reflex shows no prediction or learning, and is an example of where knowledge obtained by the systems approach can be integrated into our broad understanding of the neural basis of the pupil reflex.

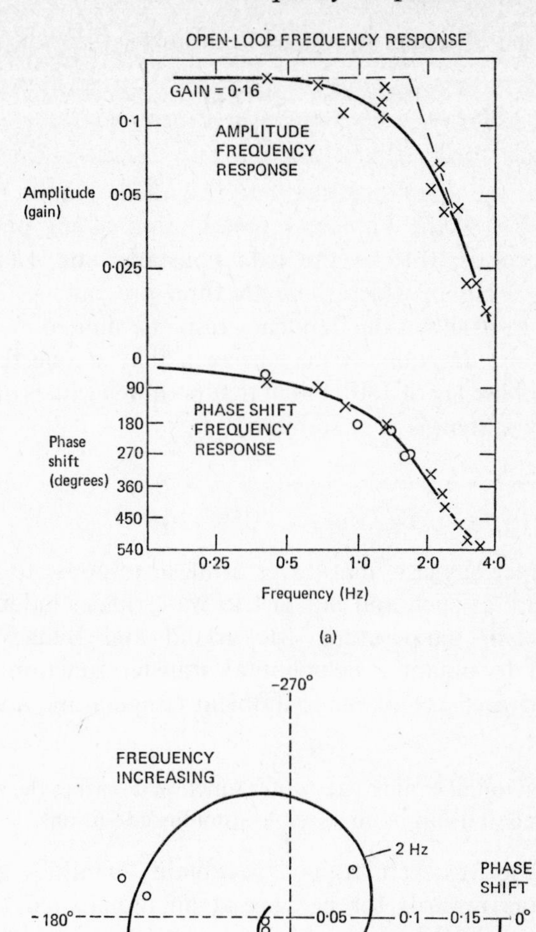

Figure 75 *Frequency response of the pupil light reflex. (a) Bode diagram, (b) Nyquist diagram. (Source: Stark and Sherman,* 1957)

Stark and Sherman produced the transfer function

$$G(s) = \frac{0{\cdot}16 \exp(-0{\cdot}18\ s)}{(1+0{\cdot}1\ s)^3}$$

to describe the open loop characteristic of the pupil. The term exp (−0·18 s) is the Laplace representation of the pure delay of 0·18 seconds, 0·16 is the gain constant, and 1/(1+0·1 s) represents a lag, of which there are three present.

Figure 75(b) shows the frequency response plotted in the form of a Nyquist diagram. It may be seen that at one frequency there is a phase lag of 180°, but at that point the gain is only 0·12, and so the system is very stable.

A BEHAVIOURAL TRANSFER FUNCTION

The evidence suggests that doves drink in response to ambient temperature as such and not just to water losses induced by a high ambient temperature. McFarland and Budgell (1970) attempted to obtain a behavioural transfer function relating thirst motivated behaviour to ambient temperature, and stated that:

In the behavioural context, a transfer function describes the stimulus-response relationship in all possible stimulus conditions.

Barbary doves were trained to obtain quantities of water (0·1 cm^3) as rewards for pecking at an illuminated key in a Skinner box. A VI 2 min schedule was used. This meant that the time between water rewards was randomised, but on average 2 minutes had to elapse between one unit of water being obtained and the bird being able to obtain the next unit. The birds were 48 h water deprived at the time of the experiment, so it was the modulation of the basic rate of responding that was being measured. The ambient temperature was made to fluctuate sinusoidally between 5 °C and 25 °C. Frequencies between 0·4 and 4 cycles/h were applied, while pecks made and rewards delivered were recorded on an event recorder.

Figure 76(a) shows sample test sessions at 0·5 and 4 cycles/h. It is clear that phase lag increases and the signal becomes

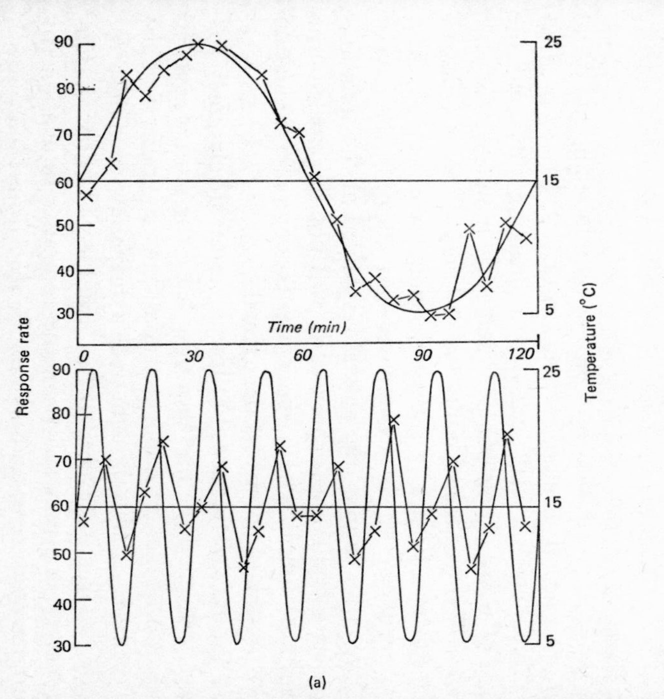

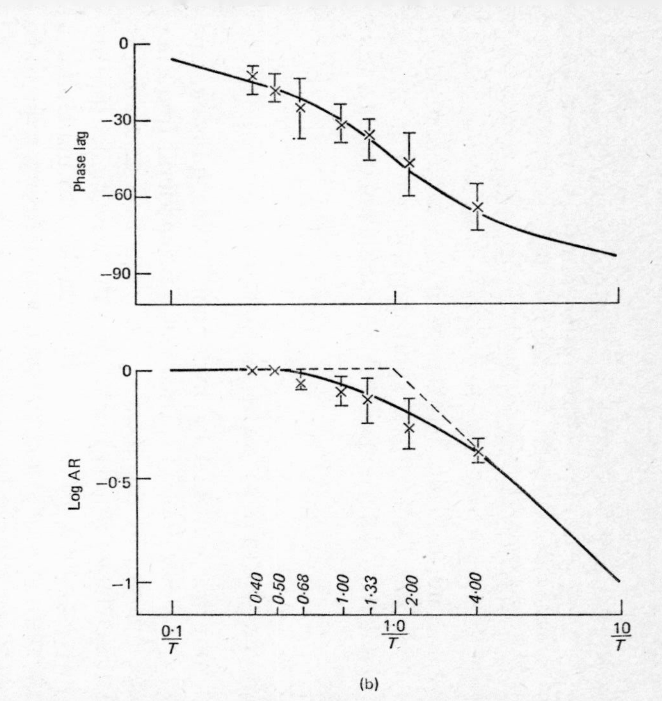

Figure 76 *The response of Barbary doves to a temperature stimulus. (a) The response at* 0·5 cycles/h *(upper) and at* 4 cycles/h *(lower). (b) The frequency response of the system plotted as a Bode diagram. (Source: McFarland and Budgell,* 1970)

attenuated as the frequency is increased. McFarland and Budgell attempted to fit the first-order transfer function

$$G(s) = \frac{K}{1+sT}$$

to the data, and, as Figure 76(b) shows, it provides a remarkably good fit. T was found to be approximately 5–6 min. It was suggested that drinking is mediated by hypothalamic temperature changes.

THIRST PRODUCED BY SALT INJECTION

Thirst may be provoked by hypertonic saline infusion, and Oatley and Toates (1971) carried out a behavioural frequency response analysis by this means. The sodium concentration in the extra-cellular compartment of rats was made to vary sinusoidally about a mean level. This was done by implanting a cannula into the rat's jugular vein, and infusing hypertonic saline at a rate defined by

$$A + A \cos \omega t$$

Thus the rate of infusion was zero at its minimum, and at no time took a negative value. Frequencies of 1, 2, 4 and 8 cycles/h were used. The rat had access to a drinking tube and the amount of water drunk was recorded every minute.

Figure 77 shows the stimulus and response for a frequency of 2 cycles/h, and Figure 78 shows the same for 8 cycles/h. Also shown are the results of a computer simulation. In the case of the 2 cycles/h stimulus, the animal responded at the same frequency as the stimulus. This clearly was not the case for the 8 cycles/h stimulus. At both frequencies the computer simulation behaved in the same way as the animal.

The failure of frequency-following at relatively high frequencies indicated the presence of at least one non-linearity in the system. In fact an important non-linear component of the thirst control system is thought to be a threshold which the error in body fluids must exceed before drinking is initiated. This non-linearity was removed from the computer simulation (by making its value zero) and, as may be seen from Figure 78(d),

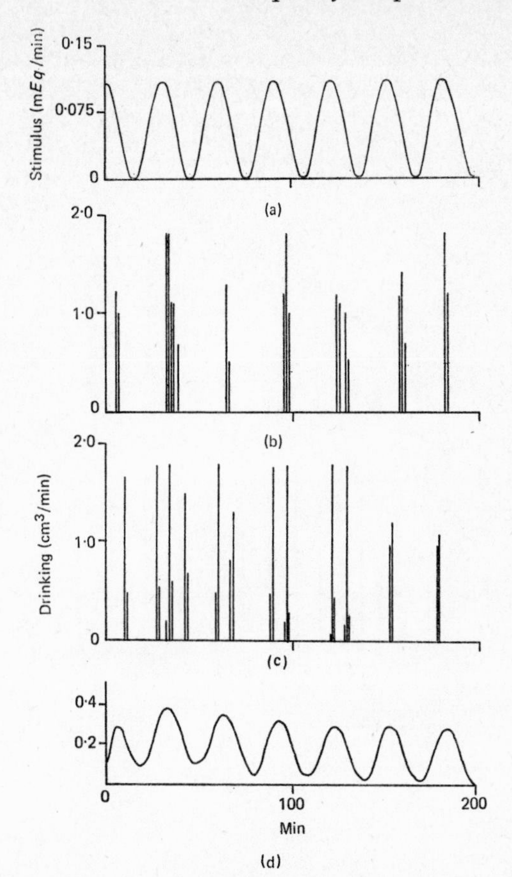

Figure 77 Frequency response of the thirst control system at a frequency of 2 cycles/h. (*a*) *The stimulus in the form of a sinusoidally varying hypertonic saline injection,* (*b*) *the response of the animal,* (*c*) *the response of the computer simulation,* (*d*) *the response of the simulation after removal of the thirst threshold.* (*Source: Oatley and Toates,* 1971)

the simulation responded at 8 cycles/h. Note that the simulated response without the threshold is quite different from either that with the threshold present, or with the behaviour of the rat.

At 1 and 2 cycles/h the animal's behaviour showed that the integral of the stimulus over one cycle was sufficient to push the error signal above threshold and initiate drinking. Whereas at 4 and 8 cycles/h the amount of saline delivered during one cycle was insufficient to traverse the area within the threshold,

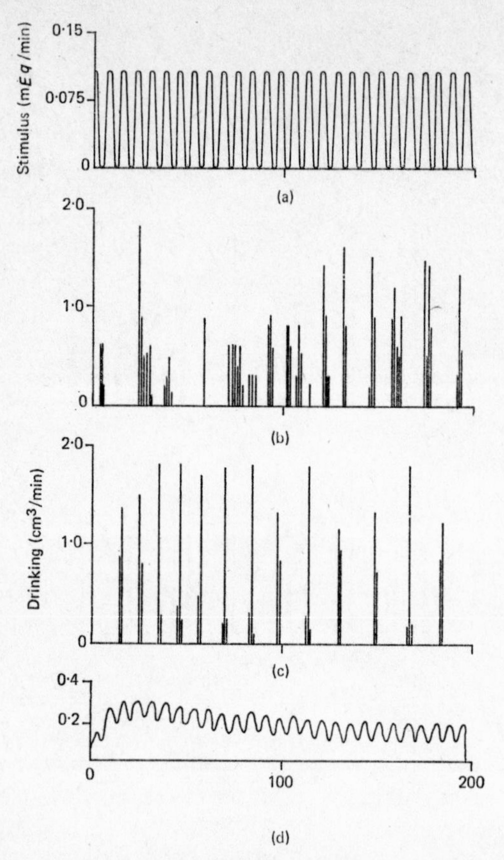

Figure 78 Frequency response of the thirst control system at a frequency of 8 cycles/h. (*a*) *The stimulus in the form of a sinusoidally varying hypertonic saline injection,* (*b*) *the response of the animal,* (*c*) *the response of the computer simulation,* (*d*) *the response of the simulation after removal of the thirst threshold.* (*Source: Oatley and Toates*, 1971)

and consequently drinking bouts occurred less often than once per cycle.

If an attempt to apply linear control theory techniques to this system had been made, some of its essential features would have been disguised and probably a quite misleading interpretation obtained.

Since a linear system responds sinusoidally to a sinewave input, when the response to a sinewave is other than sinusoidal this immediately draws attention to non-linear features of the system, and is perhaps a good reason for carrying out a frequency response analysis. Non-linear features may sometimes be exposed by comparing the system's response to both small and large amplitude sinewave inputs. This experiment forms an example of where there is a two-way flow of information between theory and experimentation. A model is constructed and then its adequacy tested against experimental results.

8 APPLICATION OF SIMULATION TO PHYSIOLOGY AND BEHAVIOUR

Physiological psychology has set itself the task of attempting to explain behaviour in terms of the physiological structure that is responsible for generating it. There are several methods by which the problem is approached. Possible chemical changes in the brain that may be correlated with behavioural measures of learning are sought. Another line of research, but one that is somewhat controversial, is that of lesion studies where anatomical locations are associated with behaviour patterns, on the grounds that a particular behavioural act suffers as a result of the loss of a specific brain region. Other researchers attempt, by means of neurophysiological recordings, to obtain a neural correlate of particular features of behaviour. In this chapter we are concerned with just one aspect of physiological psychology and that is the understanding of motivated behaviour in terms of known characteristics of physiological structure. In particular we are concerned with the explanation and simulation of homeostatic behaviour. However, this does not mean that the principles employed cannot be applied to other systems.

Homeostasis was discussed in Chapter 5, where it was noted that negative-feedback controllers serve to maintain a wide variety of bodily functions within narrow limits. The principles of homeostasis are of fundamental concern to the physiologist, but at the same time, due to the need to anchor explanations of behaviour onto secure ground, they occupy the attention of the experimental psychologist. In fact the traditional barriers between the sciences are being broken down, and one finds physiologists concerned with motivation as a means of measuring a physiological state, while psychology departments investigate the anatomy of the hypothalamus to find a neural basis for motivated behaviour. This overlap of interests is echoed in the popularity of journals such as *Physiology and Behaviour* and *Journal of Comparative and Physiological Psychology*.

The reason that homeostasis is of such interest in psychology is that a behavioural act forms a component of several homeostatic control loops – for example, drinking, eating and some forms of temperature regulation. Hunger and thirst are the two primary motivations studied by the psychologist, and presumably they arise because of an error between a reference value or system set-point and an actual value of the body's energy reserve and fluid volume, respectively. The error is the stimulus for the appropriate behaviour in each case.

Hull (1943) constructed a drive reduction theory of behaviour that has had a powerful influence in psychology, and which is based upon negative-feedback principles. Hull recognised the need of the animal to maintain optimum conditions in its internal environment, and proposed that a drive state arises when a deviation from optimum occurs. Satiation or drive reduction corresponds to reduction of the error signal to zero.

Though such a mechanism is intuitively attractive it seems to require the additional postulate that motivation can be switched off by factors other than a correction of the basic need that produced the motivation. In most cases consummatory behaviour has ended before such correction has had time to occur. Thus, if we propose that drinking is turned on by a depletion of the animal's body fluids, it is difficult to account for the cessation of drinking in terms of a restoration of the body fluids to normal. Another factor must be involved because, when drinking comes to an end, rather little of the amount drunk will have reached the body fluids; most will be in the gut.

It has been proposed that negative feedback loops convey information about the consummatory response and its immediate consequences to the brain regions responsible for generating the behaviour concerned. In the case of thirst such feedback could take the form of neural signals that are initiated by water on the tongue or in the stomach. As far as hunger is concerned, chemical messengers could possibly be employed as well as neural circuits. In this way motivation is switched on by the appearance of an error between a set-point and an actual value, but may be switched off either by local inhibitory loops sending information on the consequences of consummatory responses or by the correction of the deficit. The latter will of course occur in the long term, as the ingested substance has had time to leave the gut.

Psychologists generally speak of short-term satiety, meaning satiety caused by the presence of substances in the gut or their passage through the mouth and throat, and long-term satiety, meaning a change in the level of the physiological variable that gave rise to the motivated behaviour. Under normal conditions it would be expected that at first short-term and then long-term satiety would be experienced.

Though it is possible to speak of the various influences that contribute to motivated behaviour, in practice the psychologist usually has to be content to specify hunger and thirst merely in terms of so many hours of food or water deprivation. As a means of holding experimental conditions constant this may be adequate. However, if we want to examine the basis of motivated behaviour we must look for physiological changes within the animal that occur in response to deprivation. Questions must be asked such as: 'Can thirst motivation be explained in terms of excitatory osmoreceptors and inhibitory stomach and mouth receptors?' The theme of this chapter is that a computer simulation, based upon identifiable physiological aspects of the real system, is the appropriate tool for understanding the relation between the physiological system and the observed behaviour. However, it is good to be reminded of Milhorn's (1966) warning when we are trying to construct models:

> . . . in applying control theory to physiological systems the researcher should be certain that he is doing just this and not the opposite, i.e. attempting to apply physiological systems to control theory.

Computer Simulation

In attempting to explain a complex and often inaccessible system such as a homeostatic regulator, we require some kind of working model that incorporates our understanding of the real system. A description in words is quite inadequate since we may well be speaking in a series of contradictions. Words can disguise inconsistencies in our logic, but a working model based upon our assumptions of how the real system works will expose any faults in our theorising.

Let us consider the question of why, following water intake, an animal stops drinking when it does. In order to find an answer we must have a quantitative understanding of the changes that are occurring during drinking. Water will be leaving the stomach and entering the intestine, while at the same time water is being exchanged between the intestine and the blood, and between the extracellular fluid and the cells. It has been proposed that water in the gut inhibits thirst (see, for example, Oatley, 1967), whereas an alternative view is that a restoration of the body fluid compartments to normal is the necessary condition to terminate drinking (Corbit, 1969). Whatever our view of the satiety mechanism, it ought to be compatible with the behaviour of a model of the system that is based upon established physiological results.

In order to model a complex system such as the thirst controller we need a means of calculating, updating and exhibiting many variables at the same time. The human brain is inadequate for this purpose, and there is really no alternative to building a computer simulation. Having constructed the model, we then see what the overall behaviour of the system is like. If there is a discrepancy between this and the behaviour of the real system it indicates that either some of the basic assumptions are wrong or that we have omitted something.

Feedback systems are such that without a simulation the behaviour of the overall system may be extremely difficult to predict from an examination of the individual physiological components that comprise it. Complex and unexpected performance characteristics can arise from the subtle interactions between components. For example, a control system may exhibit oscillation that arises from properties of the closed-loop, and this might be quite unpredictable from examining the individual components. Cheyne–Stokes respiration (Milhorn, 1966) is a particularly good example of this.

The computer that we employ for our simulation can be one of two basic kinds: analogue or digital. In Chapter 6 it was shown that an analogue model of a system may easily be constructed from the block diagram. The same is true of the digital computer, though the correspondence between the block diagram and the simulation may not be so obvious.

In the analogue computer each variable is represented by an electric voltage. These may be added, subtracted, divided or integrated to give other variables. For example, the equation defining volume is given by

$$\text{volume} = \left(\int_0^t \text{flow} \times dt\right) + \text{initial volume}$$

In the analogue computer an electric voltage would represent flow, and this would form the input to an electronic integrator. Volume would be the output. For an analogue computer simulation of the body fluid and thirst system the reader can consult the paper of Reeve and Kulhanek (1967).

A digital computer simulation of the thirst–body fluid system was presented by Toates and Oatley (1970) and will be discussed here. As a starting point we can consider how we would simulate, on the digital computer, the equation just given for flow and volume. There are simulation techniques available that enable a computer programme to be written directly from the transfer function, but they are not employed here. The present simulation employs the simpler method of successive addition. It was explained in Chapter 2 that the approximate integral of a function may be obtained by multiplying the mean value of the function of time by a short time interval and then adding all such sub-integrals together. The method of successive addition, employed here, consists of examining all rate terms and forming the integral over, say, every one second of real time. Volume is therefore updated every one second.

To give an example, let us assume that the term 'flow' in the equation just given means net flow, there being a flow in and a flow out of the container. Flow out is proportional to volume and therefore the system is as shown in Figure 27 on page 80. At time zero the container holds 10 cm^3 of fluid, this being the initial conditions. The flow in is a constant, 1·0 cm^3/min. A digital computer programme to represent the system would in essence be as follows:

```
VOLUME := 10·0; FLOW IN := 1·0;
C2: FLOW OUT := K × VOLUME;
VOLUME := VOLUME + (FLOW IN – FLOW
                                OUT) × T;
'GO TO' C2;
```

T represents the time interval in minutes of real time between each calculation, and would of course be given some numerical value. To those not familiar with digital computing the expression to calculate volume probably appears somewhat odd. It is not an equation, but rather gives the specific instruction that the new volume is to be formed from the old volume plus the integral of the flow over the time interval since the last calculation was made. The programme first gives volume the value 10·0. It then calculates the flow out of the container. The integral of the net flow over the time interval *T* is calculated and added to the volume of 10·0 to give a new volume. This is one cycle complete and the 'GO TO' statement starts the programme cycling. C2 is merely a label to tell the computer where to start the cycle. It will remember FLOW IN from the start. At intervals of time *T* we obtain a measure of volume and flow out for the particular initial conditions and flow in that is specified at the beginning. As the time interval *T* is made smaller, then a more accurate solution may be obtained. The method of successive addition is not without its difficulties; it is possible under some particular circumstances to introduce instability into a system by this method. This means that the researcher must be on guard at all times, and if the problem demands it then a more sophisticated simulation technique must be employed. In the present case it was considered that the changes were sufficiently slow that the method sufficed. The simulation is in essence based upon terms with long time constants, and relatively short calculation intervals were employed.

Comparing the methods available, the analogue computer is continuous and therefore corresponds more closely to the real system when one is dealing with continuous terms such as flow and volume. However, the analogue computer has the disadvantages that drift and noise can suddenly appear in the system. Also it tends to be cumbersome to operate when a large number of components needs to be simulated.

Though the digital programme was employed for the simulation of the system, it is described here in block diagram form since this is much easier to follow. However, since the system contains numerous non-linearities it should not be thought that an analytical solution can be obtained by means of the Laplace transform.

Body Fluids and Thirst

It is necessary to convince ourselves that we are indeed dealing with a homeostatic controller when we examine drinking. Could it be that drinking is open-loop with the renal system forming the closed-loop part of the system? Some forms of drinking are almost certainly non-homeostatic, but at the same time it is clear that, in response to a body-fluid deficit, thirst will be provoked. Following a hypertonic saline injection, an animal drinks enough water to return its body fluids to normal (Corbit, 1969). After a deficit induced by deprivation, the animal's drinking is appropriate to the deficit, and after hemorrhage the amount drunk serves to replace the deficit (see Oatley, 1967). Under such conditions drinking is clearly associated with a deficit, and could not be explained in open-loop terms.

At first the simulation was constructed in order to study only homeostatic drinking, that is to say, drinking that arises from a body-fluid deficit. Other factors, such as eating dry food and circadian rhythms, influence drinking behaviour, but they appear to be inexplicable from a homeostatic point of view. They must of course be included in any full simulation – this is discussed later – but for the moment attention is directed to homeostatic drinking.

A model of homeostatic drinking, in which drinking is a function of body-fluid volume, must of necessity include a reasonable representation of the body fluids and their distribution into compartments, as well as the factors that determine body-fluid volume. Gain of water and electrolytes is via the alimentary tract and loss occurs through the kidneys and by evaporation. Simulation of homeostatic drinking therefore demands the inclusion of alimentary and renal terms. At each stage the model is built upon published physiological results.

For a detailed justification of all the assumptions built into the simulation, the reader can consult the original paper (Toates and Oatley, 1970). For ease of understanding, the simulation may be broken down into subsystems: the body fluids, stomach, intestine, renal control and drinking. Figure 79 shows the body-fluid subsystem.

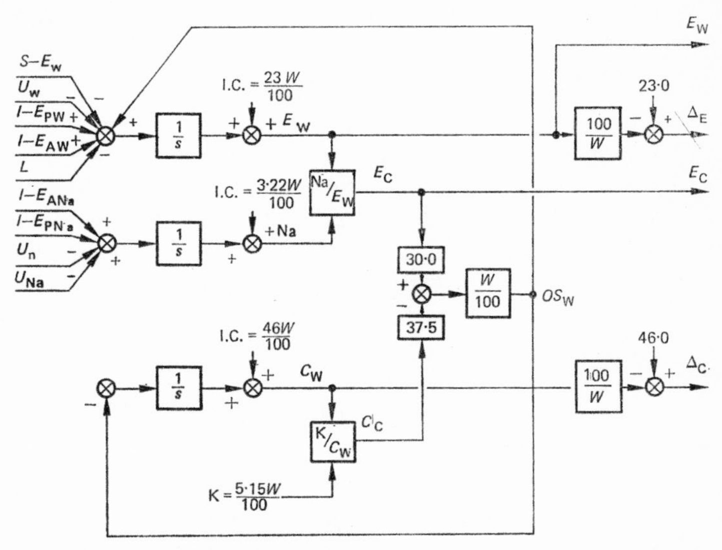

Figure 79 The body-fluid subsystem. (Source: Toates and Oatley, 1970)

The water content of the body may for our purposes be considered to be distributed into two distinct compartments: the cellular and extracellular. Sodium is largely confined to the extracellular compartment while potassium may be considered to be locked in the cellular compartment. As a first approximation the solute concentration in the extracellular space is proportional to the sodium ion concentration, and in the cellular space it is proportional to the potassium concentration. In response to a concentration gradient, water flows across the boundary between the compartments. A more detailed analysis would have to include the division of extracellular fluid between plasma and interstitial volumes. This addition to the model is in progress.

The size of the extracellular compartment of the rat is 23 cm^3/100 g body weight. For an animal of weight W the initial condition of extracellular water (E_W) is therefore 23 W/100, as shown in Figure 79. The value of E_W is given by the initial condition plus (or minus) the integral of the net flow into the compartment. Thus $1/s$ represents the process of integration of the net flow to give the change in volume. Sodium, represented by the symbol Na, is given by its initial condition plus the integral of the net movement of sodium into or out of the extracellular space. After having calculated sodium and extracellular water, we divide these (Na/E_W) in order to obtain extracellular concentration (E_C). In exactly the same way, cellular concentration (C_C) is formed by dividing potassium (K) by cellular water (C_W). To calculate the concentration gradient that causes water to move across the boundary between the two compartments, we subtract cellular concentration from extracellular concentration, this operation being performed at the summing junction. However, it is known that the normal extracellular sodium concentration of 0·14 mEq/cm^3 is in equilibrium with the normal cellular potassium concentration of 0·112 mEq/cm^3, and so clearly the concentrations must be weighted before the summing junction is reached. Equilibrium will be obtained if we multiply by the constants 30·0 and 37·5. Any two numbers in the ratio 30·0:37·5 would produce equilibrium between the normal values of concentration, but these two numbers also satisfy the dynamics of the system. If a concentration gradient appears then the size of these constants determines the flow rate of water across the compartment boundary. This models the fact that the flow rate is proportional to the concentration difference between the compartments, and the size of the constants (30·0 and 37·5) gives the best fit to experimental data for the flow rate in response to such a disturbance. The term $W/100$ following the summing junction accounts for the effect of the size of the animal on the flow rate. Presumably the cellular–extracellular boundary is twice as large for a 200 g rat as for a 100 g, and in response to any given concentration gradient the flow will be twice as large. Of course it will be twice as difficult to produce the concentration gradient in the larger animal. The term OS_W therefore represents the flow of water between the cellular and extracellular compartments.

The net flow into the extracellular compartment will be the algebraic sum of the flow of water across the stomach wall

$(S-E_W)$, the urine flow (U_W), the water exchanged between the intestine and the extracellular compartment as a result of the concentration gradient across the intestine wall $(I-E_{PW})$ and as the passenger of the active sodium flow $(I-E_{AW})$, the insensible loss (L) and the movement across the compartment boundary (OS_W). The appropriate signs are attached to each flow. Thus when extracellular concentration rises, OS_W is such as to pull water from the cellular into the extracellular compartment. OS_W has a positive sign at the extracellular summing junction and a negative sign at the cellular compartment summing junction.

The net flow of sodium into the extracellular compartment will be the algebraic sum of the active movement of sodium across the intestine wall $(I-E_{ANa})$, the passive flow across the stomach wall $(I-E_{PNa})$, the rate of urine sodium loss (U_{Na}) and the infusion rate of sodium (INF).

To the right of Figure 79 the error detectors that are employed in renal control and drinking are shown. The actual value of extracellular volume (E_W) per one hundred gram body weight is subtracted from its normal value of 23 cm³ in order to give the extracellular error signal Δ_E. Similarly the cellular volume per 100 g rat is compared to its normal value of 46·0 cm³ in order to give the error signal that arises from cellular deficit Δ_C. These two error detection operations represent the work of extra-cellular volume detectors and osmo-receptors.

Figure 80 shows the drinking subsystem. Drinking appears to be caused by a loss of water from either of the body-fluid compartments. It is worth examining in some detail the evidence

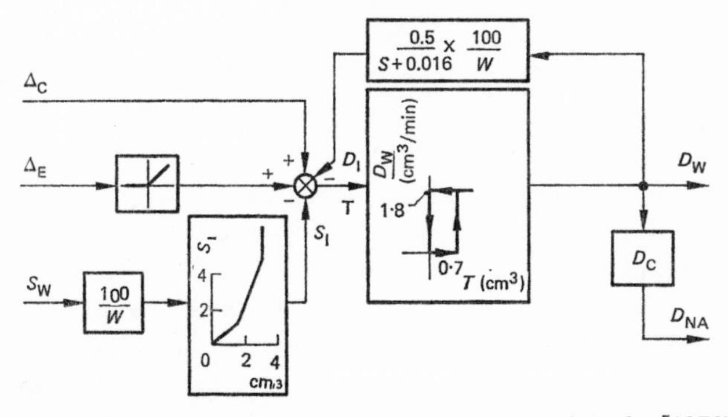

Figure 80 The drinking subsystem. (Source: Toates and Oatley, 1970)

for this assumption. Considering the cellular compartment, Gilman (1937) found that urea given intravenously was far less effective in eliciting thirst than a quantity of sodium chloride having the same osmolarity. The explanation is presumably that urea is able to cross the cellular boundary easily while sodium is confronted by an impermeable barrier. This means that water is pulled from the cellular compartment by sodium, whereas urea does not disturb the osmotic equilibrium. The implication is that loss of cellular water rather than a rise in extracellular osmolarity as such is the stimulus for thirst, though we must not rule out the possibility of a sodium receptor. Following a hypertonic injection enough water is drunk to return the cellular volume to normal. It is believed that an osmo-receptor located in the hypothalamus gives a measure of cellular volume, and when the osmo-receptor shrinks, thirst is experienced.

When the body is sodium depleted, either by means of removing the ion from the blood or by allowing only a sodium free diet for a period of time, a considerable increase in water intake occurs (Cizek, 1961; Darrow and Yannet, 1935). In response to removal of the sodium ion there is a lowering of extracellular osmotic pressure and as a consequence of this an increase in cellular water. Under these conditions thirst obviously cannot be explained in terms of loss of cellular water.

Almost forty years ago McCance (1936) in a study of the salt-depletion syndrome in humans reported that a compromise is established between the need to maintain osmotic pressure at the expense of the circulatory system and the need to maintain extracellular volume at the expense of osmolarity. Some kind of extracellular volume measure is implied.

A. V. Wolf (1958) proposed that thirst produced by sodium depletion is caused by the expansion of cellular water volume. Since hypertonic thirst is caused by cellular shrinkage, Wolf believed that a departure from normal in the opposite direction would produce a comparable neural excitation. However, as Stricker and G. Wolf (1969) pointed out, A. V. Wolf's suggestion cannot be true since it is a positive feedback loop that he describes. All that would be necessary to produce unlimited thirst would be to load with water an animal already in water balance.

The only explanation appears to be that extracellular volume reduction is a stimulus to thirst. Cizek *et al.* (1951) reported that water intake bears a relation to plasma volume reduction. Fitzsimons (1961) advanced the explanation that a reduction in extracellular sodium concentration rather than a reduction in volume is the thirst stimulus during sodium depletion. This was because McCance (1936) had shown that for humans relief could be obtained by administration of sodium chloride before any water was made available. However, this is not inconsistent with extracellular volume depletion being the prime factor. Sodium chloride would cause a shift of water from the expanded cellular space to the depleted extracellular space, and thirst would be relieved in this way.

An alternative means of producing an extracellular deficit, presumably with no change in osmolarity, is to remove blood from the body. Cannon (1932), as a result of experience in World War I, reported that loss of blood causes thirst. In controlled experiments on rats, Fitzsimons (1961) and Oatley (1964) confirmed that hemorrhage leads to water intake. It was suggested by Stricker (1966) that thirst accompanying vascular volume reduction is mediated via volume or pressure receptors. Stricker pointed out that receptors of this kind are assumed to play an important role in renal control (Gauer and Henry, 1963; Smith, 1957).

It is therefore assumed in the model presented here that a measure of both cellular and extracellular volume is made. Concerning the interaction between cellular and extracellular thirst, Novin (1962) showed that the amount of water drunk in response to a deficit was greater following deprivation than after the same hyperosmolarity caused by a salt injection. The implication of this is that extracellular hypo-volemia which occurs during deprivation adds to the effect of the cellular stimulus. Shortly afterwards Oatley (1964, 1965) showed that hemorrhage and hypertonic saline injections are additive in their effect on drinking, a result confirmed by Fitzsimons and Oatley (1968). Thus in the model Δ_C (the cellular deficit) and Δ_E (the extracellular deficit) add in order to give the total deficit signal.

Oatley (1967) proposed a qualitative model incorporating such additivity, but with the qualification that a surplus in one compartment does not inhibit the excitatory effect of a deficit in the other. The evidence that Oatley employed was that isotonic loads in the extracellular compartment are without effect on the drinking produced by hypertonic saline injections (Fitzsimons, 1961; Corbit, 1965). Another piece of evidence that could have been used to support the model is the fact that hypertonic saline injections are able to cause thirst, since any cellular deficit is accompanied by an equal and opposite extracellular surplus.

The situation regarding any possible effect of a cellular expansion on an extracellular deficit is not yet completely understood. Figure 80 shows that extracellular signals are passed only if they indicate a deficit and are blocked if they indicate a surplus. A gate element performs this function. No gate function is shown as far as cellular errors are concerned. This should not be taken to mean that no such element exists, but rather that its exact nature is still under investigation. Somehow salt depletion studies that suggest a gate element exists must be reconciled with some of Stricker's work (see Toates and Oatley, 1970) that suggests cancellation can occur.

In Figure 80, T is the thirst signal measured in cm^3, and this forms the input to a threshold hysteresis term that determines the relationship between T and the rate of drinking D_W. This operator shows that drinking rate is zero until the deficit exceeds 0·7 cm^3/100 g body weight. At this point the drinking rate takes the value of 1·8 cm^3/min. Drinking is therefore 'all or none'. The approximate exponential drinking curve following a disturbance may be accounted for by the longer pauses from drinking towards the end of the session. Drinking continues at the rate of 1·8 cm^3/min until T reaches zero, at which point D_W returns to zero. The threshold element represents the fact that the deficit must reach a certain minimum value before drinking is initiated, but once this point is passed drinking then takes thirst to zero. The choice of the expression 'takes thirst to zero' rather than 'takes deficit to zero' is deliberate and specific, the reason being the inclusion of short term feedback, to which the discussion now turns.

A subject that we have already mentioned is the fact that drinking may be halted by factors other than the restoration of body water to normal. O'Kelly, Falk and Flint (1958) showed that the time constant for the absorption of water from the gut after stomach loads is of the order of 30 min. Cellular equilibrium takes up further time. However, drinking is normally completed in a few minutes, implying some kind of short-term switch-off.

Adolph (1950), working with rats and other species, showed that the presence of water in the stomach has an inhibitory effect on drinking. Holmes and Gregersen (1950) demonstrated that water introduced into the stomach of a dog 15 to 20 minutes before a salt injection serves to inhibit drinking even though it does not prevent a rise in serum sodium concentration.

Water administered by stomach tube gives an immediate satiation though its satiety value is less than that of water ingested orally (Miller, Sampliner and Woodrow, 1957), which suggests an oral as well as a stomach switch-off mechanism. That mouth receptors can inhibit thirst is shown by the fact that animals with the oesophagus cut experience temporary satiety. The amount drunk, although larger than normal, bears a relation to the deficit.

The model therefore incorporates inhibitory loops that arise from the presence of water in the stomach and from the passage of water through the mouth and throat. S_W is stomach water; after multiplication by $100/W$ this is per 100 g rat. There is a non-linear relationship between the stomach volume and the inhibitory effect that it has – large stomach volumes have a proportionally greater effect than small ones and may even inhibit drinking completely.

Since the animal with the oesophagus cut obtains only temporary satiety from drinking, mouth and/or throat inhibition appears to be analogous to a container with a hole in it. In other words, the integral of drinking rate is formed but the integral is made to decay. The expression

$$\frac{0{\cdot}5}{s+0{\cdot}016}\times\frac{100}{W}$$

represents the operation of integration together with a decay. It is formed from the transfer function of integration $1/s$ multiplied by the exponential decay $0{\cdot}5s/(s+0{\cdot}016)$. The term $100/W$ accounts for the effect of body weight.

At the summing junction the thirst signal T is the algebraic sum of the excitatory influences arising from cellular and extracellular deficits and the inhibitory influences from stomach (S_I) and mouth and throat receptors (D_I). The tendency to drink can therefore be reduced not only by the reduction of an excitatory signal but also by the increase of an inhibitory one. Drinking will be halted when T is brought to zero, but in all probability the deficit will not be zero when the first pause in drinking occurs. If the gain of the short-term feedback loops is greater than unity, drinking may be suspended even though a deficit still exists, and it is suggested here that this is the explanation for pauses in drinking. Before long the short-term feedback effects decay, and if body fluids have not by that time been restored to normal, T is able to reach threshold again which produces another drinking bout. A series of such cycles can occur before drinking is finally halted.

With respect to the use and application of a model, there appear to be two main areas in which the simulation may be employed (Toates, 1974b):

1. Purely theoretical propositions in areas inaccessible to experimentation may be tested on the simulation to examine whether they are plausible or not.
2. A direct comparison may be made between an experimental procedure and the simulation of that procedure. This can have one of two effects. It can expose weaknesses in the model, but it can also serve to explain results that might otherwise be shrouded in mystery.

As an example in the first category, there is some controversy as to whether a threshold is or is not a component of the thirst controller (Corbit, 1969; Toates and Oatley, 1970; Oatley and Toates, 1971). Those that claim a threshold to be present could not reasonably be asked to demonstrate it anatomically in the way that a muscle could be revealed. This means that the issue is at a theoretical level, and a computer model is the appropriate tool to be employed. If the simulation is run both with and without a threshold present, it is possible to observe the predicted behaviour in each case. One is then in a position to infer whether or not the observed behaviour can best be explained in terms that include a threshold. An example of such a comparison between theory and experimentation was shown in Figure 77 (page 217) and Figure 78, where it was observed that the simulation of frequency response results with the threshold present gives the best fit.

Another experimental result that lends support to the inclusion of a threshold (Toates, 1974b) is that of Corbit (1965). Corbit found that if intravenous hypertonic saline infusions are given to rats the amount of saline that has to be injected to produce drinking is a function of the concentration of the infusion. The higher the concentration the greater was the quantity of salt needed to cause drinking.

This experiment was 'performed' on the computer model and exactly the effect noted by Corbit was found to occur (Toates, 1974b). The explanation that arises is that for low concentration infusions the increase in concentration in the extracellular fluids is slow, and consequently cellular and extracellular compartments are almost in equilibrium at all times. Thus the amount of saline present at the time drinking is initiated is just enough to exceed threshold. By contrast, for high concentrations equilibrium lags behind the rise in extracellular concentration, with the result that at the time the threshold is exceeded considerably more saline is in the extracellular compartment than would be needed if the infusion were at a lower rate. Cellular dehydration continues after drinking has been initiated. The result is dependent upon two factors: the time constant of water distribution between cellular and extracellular compartments, and the drinking threshold.

Perhaps it is as well to clarify the validity of an explanation of the behaviour of a biological system inferred from a model. If the simulation embodies the correct physiological components, then clearly explanations that arise from the interaction of these components must also be correct. It is always possible to suggest that the construction of a model is wrong, and indeed models are a fruitful source of controversy. However, this then forces new physiological results and interpretations to be obtained. The explanation is as valid as the assumptions upon which it is based.

In support of the inclusion of a threshold, Toates (1971, 1974b) proposed, on the basis of the computer simulation, that the beginning of the pauses in drinking correspond to when thirst (but not necessarily the fluid deficit) is brought to zero. The end of the pause corresponds with the thirst signal again reaching threshold. With regard to such a mechanism, it is only fair to add that other possible explanations may be advanced (i.e. the theory of McFarland (1971) that is discussed later). If the explanation just given is correct, then it says something rather specific about the location of the threshold. That is, it must be at the location shown in Figure 80, after rather than before the summing junction. If it were before, then although cellular shrinkage would have to reach the threshold value before drinking would be initiated, the long pauses in drinking could not be explained in terms of the time taken to traverse the threshold. The animal would drink until short-term effects balanced the deficit. If the short-term effect were then to decay the animal would immediately start drinking again. The rate would of course be limited but drinking would appear to be almost continuous.

What purpose could a threshold serve? If there were no threshold (and if we ignore for the moment competing motivations), the animal would always be thirsty and seeking water unless it overdrank at each session. This would not represent a good economy, since a certain error is tolerable and the threshold means that less demand is placed on the animal.

A threshold in the case of the sensory system arises because a signal has to be detected when it is buried in noise. This interpretation probably cannot be applied to thirst since the system can measure subthreshold signals, as is shown by the fact that the body deficit is returned to zero. It could be that the error can only take command of behaviour when error exceeds threshold.

Figures 81 and 82 show further comparisons between the computer simulation and experimental results. In Figure 81 it may be seen that in response to a continuous infusion of hypertonic saline both the real and the simulated rat drink in bursts of 1–3 cm^3 with pauses of 10–30 min between bursts. In the model such intermittency is dependent upon the threshold.

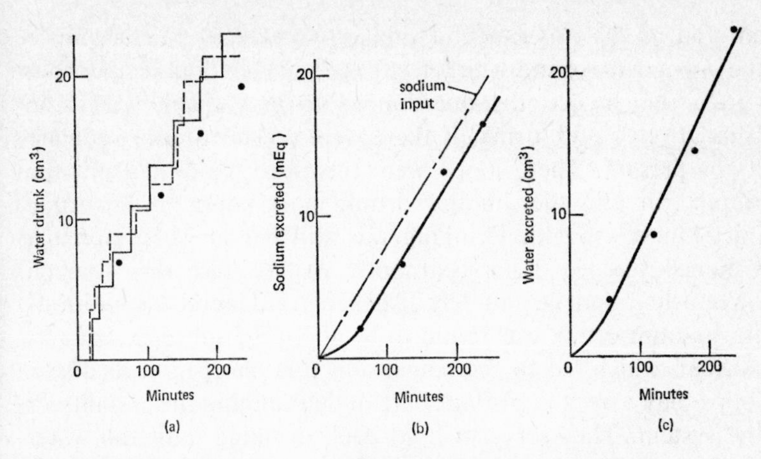

Figure 81 The response to a continuous intravenous saline infusion at a rate of 0·0405 mEq *of* Na *per minute for a* 430 g *rat*. (*a*) *Drinking*, (*b*) *sodium excreted in the urine*, (*c*) *water excreted as urine* (——— *computer simulation*, ----- *typical experimental result*, ● *mean experimental result*). (*Source: Toates and Oatley*, 1970)

Figure 82 illustrates that in response to a sudden injection of hypertonic saline the simulation 'drinks' approximately the same

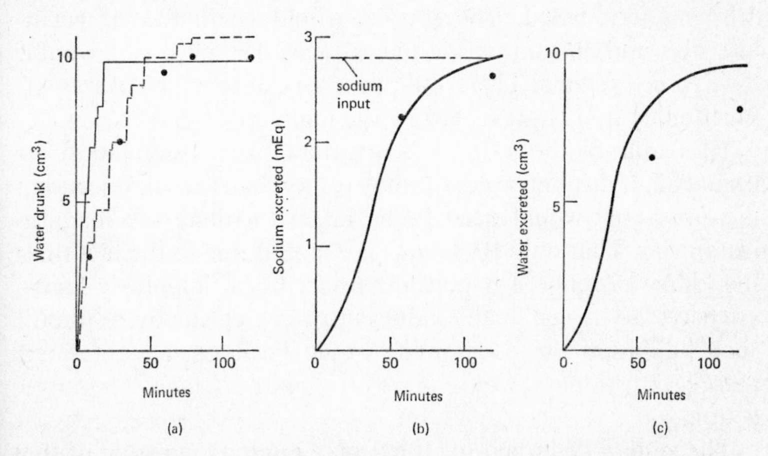

Figure 82 The response of a 430 g *rat to a step intravenous saline injection of* 2·73 mEq *of sodium*. (*a*) *Drinking*, (*b*) *sodium excreted in the urine*, (*c*) *water excreted as urine* (——— *computer simulation*, ----- *typical experimental result*, ● *mean experimental result*). (*Source: Toates and Oatley*, 1970)

amount as the experimental subject, namely 10–11 cm^3. Since the importance of short-term feedback has been doubted (Corbit, 1969), this invited the question as to how much water the simulation would 'drink' if there were no short-term feedback loops present. These loops were therefore removed from the simulation and the amount drunk in response to the same injection was noted. Drinking was still halted when stomach volume exceeded 3 per cent body weight since the stomach obviously cannot be infinitely distensible. Under these conditions the amount drunk was found to be about 3·5 times as much as normal, which led to the conclusion that short-term feedback loops play a very important part in determining the stability of the system. They serve to hold back drinking until the water reaches the body fluids. This is done by informing the brain regions responsible for thirst that correction of the deficit is on the way and therefore no more corrective action is called for. If there were no such loops, after the rat had drunk the necessary 10 cm^3 most of the water would still be in the alimentary tract. The deficit would not have been eliminated, and therefore the animal would continue drinking. Considerably more than 10 cm^3 would be taken before the deficit could be corrected and drinking terminated. The surplus would be in the stomach, intestine, and, in some cases, the extracellular space, and would have to be excreted. This would be a very uneconomical way of functioning and is prevented by the inhibitory loops.

The value of the kidneys is apparent from examination of Figure 82. If drinking were the only response to the salt injection, then 19·5 cm^3 would need to be taken to dilute the load to isotonicity. That only 10–11 cm^3 is drunk is due to the fact that the kidneys excrete a hypertonic urine. Local inhibitory loops are necessary to enable the kidneys to work efficiently. Without such inhibition the body fluids would be flooded with water before the kidneys would have a chance to take corrective action.

The system described for thirst may remind the reader of the temperature control system of the body that was discussed in Chapter 5. In both cases a central function is being regulated: body fluid content and body temperature. In both cases there are long delays in the system that produce a tendency to overshoot and instability, but the delays are compensated for by having bypass loops that quickly convey information from the periphery to the centre of the system.

It is important to emphasise that in the model a combination of both a partial return to normal of the body fluids and a peripheral inhibitory signal acting together serve to terminate drinking. Therefore it cannot be said that either central or peripheral factors are exclusively responsible. In this context, Hatton and Bennet (1970) reported that the stomach contents remained constant during the two minutes before and after drinking was terminated. They concluded that such a result casts doubt on the importance of stomach distension as an inhibitory factor in thirst. According to the interpretation presented here the result points to the importance of other factors in satiety, such as cellular uptake of water, but it does not diminish the importance of stomach distension. If a stomach load of water diminishes thirst, then logically water in the stomach as a result of drinking must also inhibit drinking.

PHYSIOLOGICAL AND BEHAVIOURAL MODELS

Construction of a model of a behavioural control system from basic physiological components raises the question of how far one can explain behaviour purely in terms of physiologically recognisable terms such as fluid deficits and inhibitory signals from the stomach to the thirst regions of the brain (Toates, 1974b). For a start it must be acknowledged that the simulation has nothing to say about non-homeostatic drinking, such as schedule-induced polydipsia (Falk, 1961) or frustration-induced drinking (Panksepp, Toates and Oatley, 1972). In such cases the amount drunk bears absolutely no relation to any fluid deficit, and indeed drinking can occur even though the body fluids are seriously overloaded. But can the model provide a reasonable account of homeostatically based behaviour?

McFarland (1971) holds the view that although the model of Toates and Oatley (1970) can go a long way towards explaining thirst behaviour, there is a feature of behaviour about which the model has nothing to say. To give an example, as may be seen from Figure 82, in response to an injection of hypertonic saline the simulation 'drinks' at a faster rate than the animal. McFarland's explanation for this disparity is that the computer simulation concentrates on the one activity of drinking, whereas in reality the rat's time will always be shared with other activities such as grooming. According to McFarland, even in the presence of a dominant thirst motivation, attention will still be switched at intervals to other behavioural channels further down in the hierarchy. It should be noted that in the long-term there is quite good agreement between the simulation and the experimental result. Only in the speed of response is there a disparity.

The approach of McFarland and the method employed here represent two quite different ways of studying the same problem. McFarland's method pays little attention to the underlying physiological components, but rather establishes relationships between experimental stimuli and the observable behaviour of the animal. What lies between is a 'black box' exhibiting certain characteristics, one of which is time-sharing. By contrast, the model of Toates and Oatley is built upon identifiable physiological components, and in such terms cannot account for time-sharing.

There is no reason why a time-sharing device exhibiting the characteristics described by McFarland should not be built into the model of Toates and Oatley.

EFFECT OF A DELAY IN DRINKING

In an experiment carried out on Barbary doves, McFarland and McFarland (1968) allowed thirsty birds to obtain water by delivering it as reward for pecking at a key. Each unit of reinforcement was 0·1 cm^3 and 5 s had to elapse between rewards. After drinking was in progress a 5 min interval was forced during which water was unavailable, the bird being informed of this by the fact that a light was extinguished. It was found that after the interval the birds drank at a slower rate than they were immediately prior to the pause. McFarland and McFarland concluded that there is a positive-feedback loop such that the act of drinking tends to keep drinking going once it is initiated. Such a positive-feedback loop is included in McFarland's model shown in Figure 62 on page 187.

The model of Figure 80 specifically includes negative rather than positive feedback in connection with the act of drinking. As a result, this model predicts that if a pause is imposed drinking will be resumed at a faster rate after the pause.

A number of possibilities emerge:

1. There is a species difference such that there is a positive-feedback loop in the case of the Barbary dove but a negative-feedback loop in the case of the rat.
2. The experimental result obtained by McFarland is due to the special experimental conditions employed.
3. There is indeed a positive-feedback loop in the case of the rat.

Examination of Figure 77 on page 217 shows that drinking does tend to be in more concentrated bouts than would be suggested by the simulation, and this indicates some kind of inertial factor. However, negative rather than positive feedback accounts for the result that animals obtain temporary satiety when the oesophagus is cut. On the basis of McFarland's positive-feedback model, the animal with its oesophagus cut could never obtain even temporary satiety. The slightest thirst would promote unlimited drinking.

MEAL-ASSOCIATED DRINKING

Each gram of food eaten pulls approximately 1·0–1·5 cm^3 of fluid into the alimentary tract; the time course of this movement has been measured by Oatley and Toates (1969).

The physiological changes in the form of dehydration that follow eating were simulated on the digital computer (Toates, 1974b) and drinking was shown to be based upon the fluid deficit. It was found that the simulation gave a very poor fit to the rat's behaviour, which suggested that meal-associated drinking is not a consequence of the fluid deficit. On later reflection it may be seen that this accords with the experimental results. Oatley and Toates (1969) showed that the appearance of fluid in the alimentary tract following eating is describable by a time constant of 10–15 min. However, drinking appears sooner than can be accounted for in terms of dehydration; the animal drinks in anticipation of dehydration rather than in response to it.

On the basis of the computer simulation, Toates (1974b) suggested that there is an enormous value in such anticipation. If drinking were error actuated, in order for the animal to drink, fluid would have to leave the extracellular space and enter the gut. This would promote drinking which would enlarge the gut volume still further – a cumbersome way of getting water into the body fluids. If the animal drinks in anticipation of dehydration then dehydration is avoided. In the experiment of Epstein and Teitelbaum (1964) a rat that is recovering from lateral hypothalamic damage does not drink in response to stimuli of central locus, such as deprivation, salt injections, or hypovolemia. However, it does drink in conjunction with eating dry food.

This is not the first time that we have met the idea of long delays being avoided by an anticipatory action that prevents the incurrance of an error.

It is possible that additions can be made to a physiological model of thirst so that the model represents behaviour inexplicable in terms of homeostasis. In this way homeostatic and non-homeostatic behaviour may be studied in order to understand behaviour that comprises both.

INHIBITION OF HUNGER BY THIRST

During water deprivation an animal cuts down on its food intake. Since food requirements are not lessened by water deprivation, this points to some kind of inhibitory interaction from the thirst controller to the hunger controller. Oatley (1967) proposed a neural inhibitory pathway that serves to subtract a signal from the hunger signal.

A series of experiments have served to elucidate the nature of the inhibition (Toates and Oatley, 1972; Oatley and Toates, 1973). Particularly, the experiments were addressed to the questions as to whether thirst interacts subtractively or multiplicatively, and how much inhibition is produced.

A subtractive inhibition would be defined by the equation

$$F_{\mathrm{T}} = F - K_{\mathrm{T}}$$

where F_T is the amount of food eaten when the animal is thirsty, F is the food eaten by the non-thirsty animal and K_T is the inhibitory effect of a particular level of thirst, a constant for a constant thirst level.

A multiplicative interaction would be defined by the equation

$$F_T = FK_T$$

where K_T is a constant for any level of thirst and always has a value of less than unity.

In the case of subtractive inhibition the amount that the animal eats in the presence of thirst will always be less by some constant amount than that it would eat when not thirsty, and the difference will be irrespective of the level of hunger. For example, a hungry animal might normally eat 10 g and this would be reduced to 9 g when accompanied by thirst. When less hungry it might eat 4 g, and this would mean under the same hunger conditions it would eat 3 g when thirsty. In the case of a multiplicative interaction the figures might be 10 g (hungry and not thirsty), 9 g (hungry and thirsty), 4 g (slightly hungry and not thirsty) and 3·6 g (slightly hungry and thirsty).

The eating behaviour of rats was studied at various levels of hunger after salt injections (Oatley and Toates, 1973) and it was clear that thirst had a subtractive effect upon hunger. Irrespective of the level of hunger, each cm^3 of thirst inhibited about $\frac{1}{3}$ g of feeding. Allowing 1 cm^3 of water to drink had the effect of disinhibiting $\frac{1}{3}$ g of feeding, a confirmation of the inhibition figure.

It was argued that since each gram of food eaten causes about 1 cm^3 of thirst (see the last section) and that each cm^3 of thirst inhibits about $\frac{1}{3}$ g of feeding, a rat not made thirsty by deprivation or salt injection but eating food in the absence of water should eat about two-thirds as much as the animal with water. This was indeed found to be the case.

The inhibition of eating by water deprivation was also studied (Toates and Oatley, 1972), and it was found that at high levels of thirst the amount of eating inhibited by each cm^3 of deficit was larger than one-third. This suggested a non-linear relationship between thirst and its inhibitory effect, but did not of course argue against a strictly subtractive mode of interaction.

On the basis of the experimental results a model in systems fashion was proposed. This model is shown in Figure 83. A

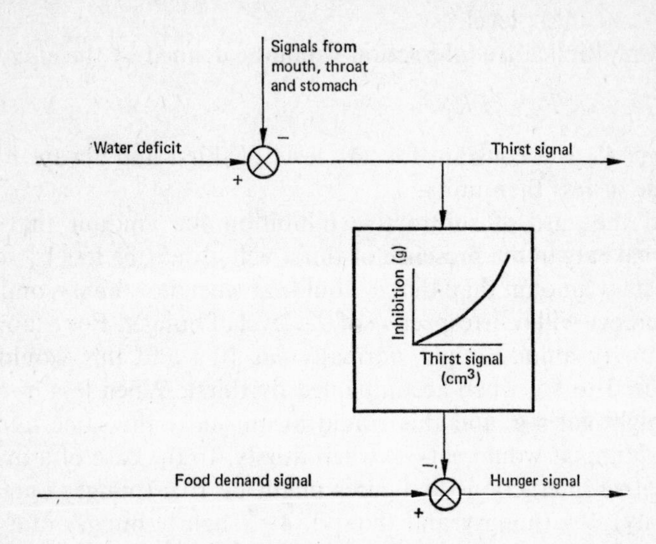

Figure 83 *A control model of the effect of thirst on hunger*. (*Source: Toates and Oatley*, 1972)

thirst signal subtracts from a food demand signal in order to give a hunger signal. In this way thirst can reduce or even completely cancel the effect of a food demand signal arising from an energy need. The relationship between thirst and its inhibitory effect is shown to be non-linear.

As has already been discussed, the total thirst signal is thought to be the algebraic sum of excitatory cellular and extracellular deficit signals and inhibitory signals from the mouth, throat and stomach. It appears also that it is this total thirst signal that inhibits feeding. Cellular and extracellular deficits are additive in their inhibitory effect on feeding (Hsaio, 1970) just as they are in their excitatory effect on drinking. The rapidity with which feeding is disinhibited following drinking indicates that short-term feedback does indeed act as shown in Figure 83. It would not be expected that osmo-receptors could respond so quickly (Oatley and Toates, 1973), though there is insufficient evidence to be certain on this point.

CIRCADIAN RHYTHMS

Understanding the circadian rhythm forms one of the most fundamental problems in biology. In the present context the circadian rhythm is revealed in the fact that typically a rat drinks about 80 per cent of its daily intake of water in the 12 h of darkness and only 20 per cent during the 12 h of light.

Oatley (1974) generated a circadian rhythm on a digital computer and then investigated various modes of coupling the diurnal rhythm to a simplified version of the homeostatic controller of drinking presented by Toates and Oatley (1970). As Oatley points out, any proposed interaction is subject to specific constraints. The drinking rhythm must be capable of showing entrainment on the lighting cycle, even if the period is changed (within limits) and if the phase is shifted. It must also be capable of maintaining oscillation in constant darkness.

One might be tempted to suggest that the circadian rhythm is something independent of the homeostatic controller, but if this were the case one cannot explain why rats sometimes drink nothing during the 12 h of light. Oatley comes to the conclusion that during this period drinking is suppressed below that dictated by homeostatic needs.

TOWARDS A PHYSIOLOGICAL CONTROL MODEL OF FEEDING

Due to the fact that the interactions from hunger to thirst are of crucial importance in understanding drinking behaviour, there is a need for a physiological model of eating, not to mention the need for such a model in the study of hunger. It is probably much less the case that successful simulation of feeding must await the appearance of a more advanced model of thirst.

An attempt to model a very first approximation to feeding in the rat was recently launched. Figure 84 shows the model.

Rate of absorption of energy from the gut is subtracted from the rate at which energy enters the gut from feeding, and the difference gives the net rate of supply of energy to the gut. The integral of this net rate gives the gut content of energy (assuming that the energy content of the gut is zero when we begin the observations). Absorption from the gut appears to be proportional to the square root of the gut content, and in the model the

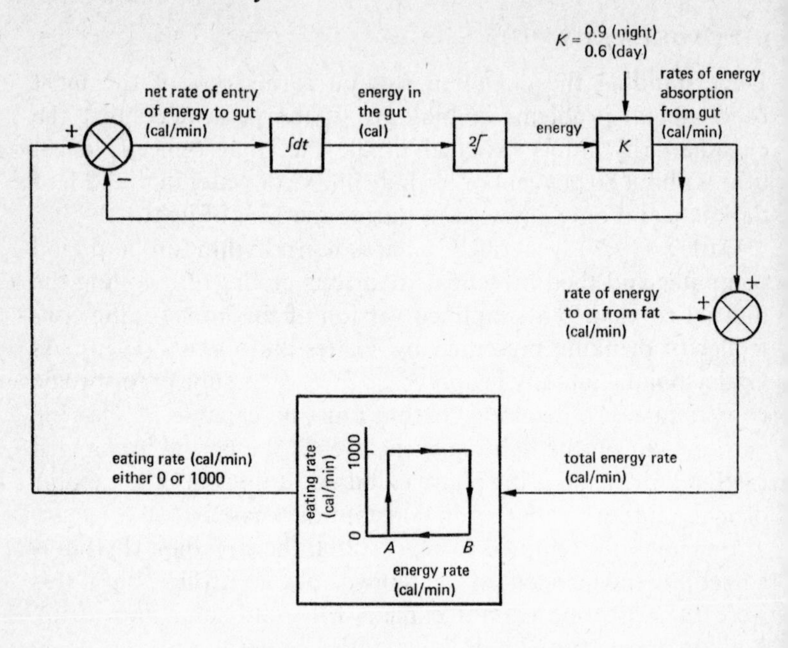

	Dark		Light	
	Model	Exp. rat	Model	Exp. rat
Normal	13·0	17·4	7·6	7·6
Dynamic hyperphagia	13·0	17·7	13·6	15·5
Static hyperphagia	10·5	9·2	7·2	7·9

Figure 84 A model of feeding in the rat. The table shows the predictions of the model for the amount eaten in grams for the two 12 h *periods, which are compared with experimental results.* (*Amer. J. Physiol*, (1971), **221**, 711. *Taken from a model of feeding produced by D. A. Booth and F. M. Toates*)

square root of the gut content is calculated and this is multiplied by the gain *K* in order to get the rate of energy absorption from the gut, measured in calories per minute. For any particular point in time *K* is a constant but it shows a circadian variation, and is larger (0·9) in the dark phase than in the light phase (0·6). In the physiological domain this means that the gut empties faster during the night than during the day.

In the model feeding may take one of two values, zero or 1000 calories of utilisable energy per minute (0·34 g of chow per minute). In other words the system is all or none; the rat either eats at this rate or it does not eat at all. Total energy rate is the supply to the system of readily metabolisable energy and is given by the sum of the energy derived from the gut and that supplied by net mobilisation and storage of lipid. The latter also shows a pronounced circadian variation; in the dark phase there is a conversion of energy to fat and in the light phase there is a release of energy from fat.

When the net energy supply rate falls to below value *A*, feeding is switched from value off to value on, and eating proceeds at a rate of 1000 cal/min until energy supply rate reaches value *B*. It is then switched to zero. Value *A* was estimated to be 18 cal/min and B to be 60 cal/min.

The model predicts meal sizes of between 2 and 3 g, the main part of the animal's intake being at night. As the table in Figure 84 shows, there is reasonable agreement between the simulation and the real rat. A simulation of the dynamic phase of ventromedial hypothalamic hyperphagia was attempted by making lipid exchange a net synthesis rate by day as well as by night. The model predicts that such a rat will take larger meals by day, and this accords with experimental results. The model also offers an explanation. The rat takes larger meals during the day than during the night because gut emptying is slower during the day. It is necessary to have a large gut content in order to get the same rate of gut absorption of energy as would be obtained with a relatively small gut volume during the night. Since gut content is greater the gut takes a longer time to empty and the meals are more widely spaced by day than by night. The same applies in the case of the normal rat but here the meal size is smaller because of the assistance of lipolysis which is absent in the lesioned rat.

General Conclusions

This chapter has presented a model of thirst that is believed to incorporate our current understanding of the underlying physiology. Additions and alterations will no doubt have to be made to the model, and indeed some of these are now in progress. No model should be looked upon as representing the final word, but rather the embodiment of our somewhat limited understanding.

An example of a problem in biological systems that has been brought to light by model building is that of what constitutes the reference value. Oatley (1967) claims that the existence of a reference value in a biological system has yet to be shown and suggests that biological systems may be self-optimising in some way. This may well be the case for some systems, but in other cases the division between an optimising system and a set-point system seems to be only academic. To take one example we may return to the controllers of eye position. Since the object of eye movements is that we should obtain single vision with the image and fovea in alignment, we can consider that the reference value of the system is the foveal position. Indeed, departure from the fovea is the cue for a corrective response. In the case of version eye movements, when the two images fall to the same side of the fovea a conjugate response results. Vergence is stimulated when there is a disjunctive disparity. Output is eye position and error is the difference between the actual position of the eyes and the required position. However, since the eye muscles drive the eyes until the position of best acuity (the foveal position) accompanied by single vision is obtained, we could look upon it as being an optimising system. The eyes are driven until the optimum position (the peak of the hill) is found. Any departure from the peak causes a corrective action; the reference value is the same as the peak of the hill.

If it is the case that the shrinkage of a particular cell in the hypothalamus causes drinking, then the volume of this cell under normal conditions assumes the properties of a reference value. It is not at all obvious that this represents an optimum value in the sense that the foveal position is clearly an optimum, but millions of years of evolution stand behind the body's homeostats, so it would not be at all surprising if it were. Cell shrinkage causes drinking and drinking causes the cell to be restored to normal. In an engineering system the normal cell size would constitute the reference value, so it is largely an academic question as to whether we are allowed to borrow the expression for a biological controller.

These arguments should not however be taken to mean that optimisation can always be written in terms of a set-point. Certain biological systems may operate by searching for an optimum along two or more dimensions. Oatley cites the work of Priban and Fincham (1965) on respiratory control, in which action appears to be taken to maintain the system on the peak of a hill relating blood chemistry to the efficiency of gaseous exchange.

The ultimate purpose of homeostatic systems is not to maintain one function constant at any price but rather to maximise the organism's chance of survival. The physiological embodiment of this is the fact that homeostatic systems do not exist in isolation, on the contrary profound interactions exist between them. Hunger and thirst represent particularly good examples of this. The homeostatic regulation of energy, involving feeding, acts against the interests of survival if the animal is water deprived, since the movement of water into the alimentary tract following food intake causes embarrassment to the circulation. Thus the fluid-balance homeostat dominates, and it does so by means of an inhibitory pathway from thirst to hunger. Model building is a good way to understand such interactions and, as a result, to understand the hierarchy of homeostats also.

Behaviour is to be understood in terms of a subtle interaction between homeostatic systems definable in control systems fashion and non-homeostatic components. Of first importance is the diurnal rhythm, a non-linear oscillation that we know very little about. To give but one example of the importance of the diurnal rhythm in biology, sleep is a phenomenon of which we have a few empirical observations but almost no understanding of the underlying mechanism. Just as biological control-systems analysis has flourished by borrowing feedback theory from engineering, so we need to adapt the theory of non-linear oscillation to biological systems and ultimately fit the two theories together.

As a final word we return to the idea, introduced in Chapter 5, that systems having very different embodiments can exhibit identical performance. In the context of discussing the set-point or reference value of a system one should be aware that something very much resembling a set-point can appear even though strictly speaking the system does not contain a set-point. The set-point concept involves a measurement of error, corrective action being taken in response to the error. There is reasonable evidence that drinking in response to body-fluid volume is an example of a system involving a set-point. Turning to the companion motivation of eating, can we find a set-point for body weight?

Something resembling a set-point appears to be present since after a period of starvation an animal's body weight increases when food is made available. Forced over-feeding causes an increase in weight, but as soon as forced feeding is ended the animal voluntarily reduces its food intake until its weight returns to normal. Does this necessarily mean that body weight or, say, body fat quantity is measured and compared with a reference value? Figure 85 shows an alternative to the set-point concept that is tentatively offered. Let us suppose that lipolysis (the rate at which the fat deposits in the body are being broken down to yield energy) is a function of the size of the fat stores in the body, as shown in Figure 85(a). This is a negative-feedback effect – the larger the size of the fat deposit the larger the rate of fat breakdown. This would occur if large adipose cells have a larger compartment of readily mobilised triglyceride. Figure 85(b) shows a relationship between size of the fat stores and lipogenesis (the rate at which energy is converted into fat), and again a negative-feedback effect is seen. A possible explanation for the shape of this function is that insulin resistance tends to develop in obesity. That is to say, with large fat deposits a given level of insulin may be less effective at inducing the glucose uptake into adipose cells which is necessary to permit lipogenesis.

In Figure 85(c) the lipogenesis and lipolysis functions are shown in the same diagram. The net rate at which energy is converted to or from fat will be the difference between the two curves. Thus when the body fat stores have the value A there is a net lipogenesis given by value a, and the body will be increasing in weight. At a body fat store value of B there will be a net lipolysis whose magnitude is given by the value b; in other words the fat deposits are being broken down. In fact life will not be as simple as Figure 85 suggests (the lipogenesis value will also depend upon the availability of food), but let us assume that we

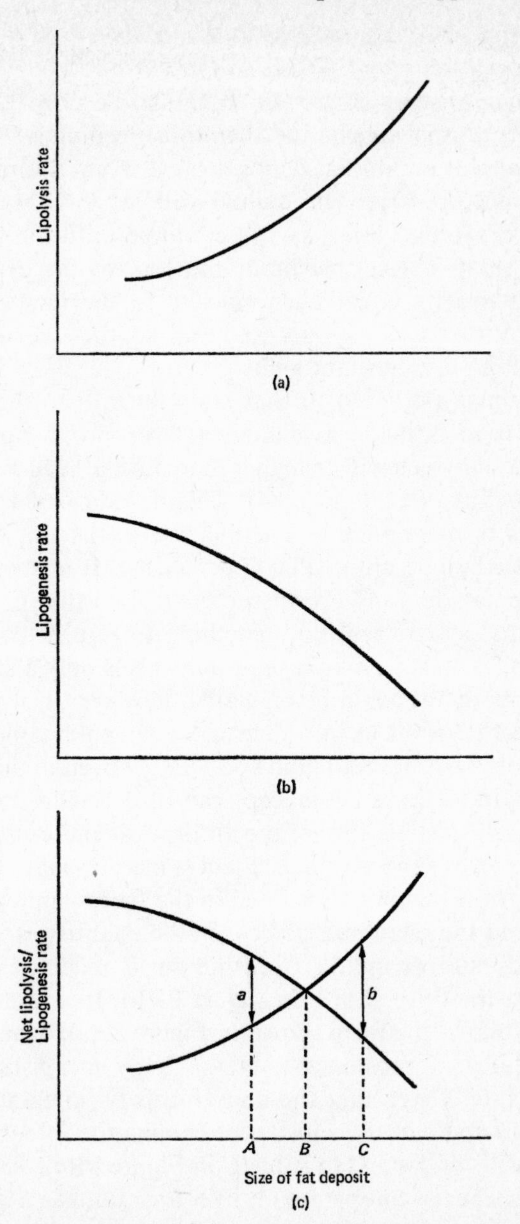

Figure 85 *The factors which contribute to the size of the fat deposit. (a) Lipolysis and (b) lipogenesis; (c) shows the net result. (Taken from a model of feeding produced by D. A. Booth and F. M. Toates)*

are observing what happens over a period when food is available ad lib.

Let us take the case of an animal that has been force-fed until it is over its normal weight and then force-feeding is halted and food is available ad lib. Fat deposits are at point B and so there will be a net lipolysis. The animal will eat less than normal (see Figure 84), and lipolysis will continue until the fat stores have the value C. At this point the loss of fat deposits by lipolysis is exactly equal and opposite to the increase in fat deposits by means of lipogenesis, and so the size of the fat deposits stays at a constant value.

If an animal is starved so that body fat falls to the value A and then food is made available, a large proportion of the calories eaten (which will be higher than normal) will be diverted to the fat stores (see Figure 84). This is indicated by the net lipogenesis of magnitude a. The increase in the size of the fat stores will continue until value C is reached. It is obvious then that if the fat deposits are disturbed from value C in either direction the system will do something to return to C if it is allowed to. It must be emphasised that this is only a suggestion as to how body fat could be regulated. However, if it proves to be incorrect it serves to introduce a fundamental concept, that of the ability of systems to home on a target even in the absence of a set-point. The same concept can also be illustrated by a quite different system, that of the absorption and excretion of a substance. We suppose that a substance x may be either absorbed from the gut or excreted in the faeces, and we further suppose that the relationship between the quantity of substance in the body and its uptake from the gut is reflected in a relationship of the form shown in Figure 85(b). Its excretion from the body follows the form shown in Figure 85(a); that is to say, as a quantity of substance x in the body increases so does excretion rate. Given that the body forms an integrator of the net rate of supply of the substance the quantity of substance in the body will correspond to value C in Figure 85(c). This type of control is believed to operate on iron for example.

Is C the set-point of the system? The answer depends upon how we define set-point. If we debase the original concept to say that it is the value that the system tends to home in on then the answer is yes. If, however, we keep to the original meaning and define set-point as that value with which the output is compared in order to extract an error signal, then C cannot qualify as set-point.

These remarks now bring my argument to an end. I have added some 50 000 words to the biological literature, and I feel in conclusion that I must justify this. Why I feel qualms of conscience is that the biological sciences are now confronted with what is known as an information explosion. The number of published articles rises as a function of t raised to some alarming power. It is difficult to see how we can handle the problem, unless we have some means of rationalisation.

Although a control model of a system such as body fluid regulation is complex, it has the enormous advantage that it enables information to be brought together and represented in an economical and logical way. In addition to anything else, it is essentially an abstracting service. It is worth exploring the possibility that models, together with the references on which they are based, could be built into a highly efficient information storage system.

Let us again return to engineering, a source that we have borrowed so much from already. The design of a complex system, such as an aircraft, is broken down into component subsystems. Diagrams show each subsystem, while other diagrams sacrifice the detail to show the interconnections between subsystems. One convention of notation is employed throughout, the design involving a rigid hierarchy of organisation. In this way an otherwise impossible task becomes manageable. If the component parts were to be described in articles lost in the engineering literature, and nowhere else could one find an account of the interactions, then the task would not be easy.

I am reluctant to make superficial comparisons that contain little useful information, but at the same time I cannot help feeling that there is much more to be gained from engineering than we have imagined. Every week hundreds of new experiments appear, and it is my belief that we need theories that can bring them together. Such theories can take the form of models. A theory is useless without experimentation to reinforce its validity, but are experiments any less dependent upon the theory?

REFERENCES

ADAMS, J. A. (1961). Human tracking behaviour. *Psychological Bulletin*, **58**, 55–79.

ADOLPH, E. F. (1950). Thirst and its inhibition in the stomach. *American Journal of Physiology*, **161**, 374–86.

ALPERN, M. (1969). Types of movement. In *The Eye*, H. Davson (ed.), vol. 3, 65–174. Academic Press.

BENZINGER, T. H. (1964). The thermal homeostasis of man. In *Homeostasis and Feedback Mechanisms*, G. M. Hughes (ed.), 49–80. Cambridge University Press.

BENZINGER, T. H. (1969). Heat regulation: homeostasis of central temperature in man. *Physiological Reviews*, **49**, 671–759.

CAMPBELL, F. W., and G. WESTHEIMER (1960). Dynamics of accommodation responses of the human eye. *Journal of Physiology (London)*, **151**, 285–95.

CANNON, W. B. (1932). *The Wisdom of the Body*. The Norton Library, New York, reissued 1963.

CASSWELL, S., and D. MARKS (1973). Cannabis induced impairment of performance of a divided attention task. *Nature* **241**, 60–1.

CIZEK, L. J. (1961). Relationship between food and water ingestion in the rabbit. *American Journal of Physiology*, **201**, 557–66.

CIZEK, L. J., R. E. SEMPLE, K. C. HUANG, and M. I. GREGERSON (1951). Effect of extravascular electrolyte depletion on water intake in dogs. *American Journal of Physiology*, **164**, 415–22.

COGAN, D. G. (1937). Accommodation and the autonomic nervous system. *Archives of Ophthalmology (NY)*, **18**, 739–66.

CORBIT, J. D. (1965). Effect of intravenous sodium chloride on drinking in the rat. *Journal of Comparative and Physiological Psychology*, **60**, 397–406.

CORBIT, J. D. (1969). Osmotic thirst: theoretical and experimental analysis. *Journal of Comparative and Physiological Psychology*, **67**, 3–14.

CRAIK, K. J. W. (1947). Theory of the human operator in control systems. *The British Journal of Psychology*, **38**, 56–61, 142–8.

CRANE, H. D. (1966). A theoretical analysis of the visual accommodation system in humans. Stanford Research Institute, Project 5454.

DARROW, D. C., and H. YANNET (1935). The changes in the distribution of body water accompanying increase and decrease in extracellular electrolyte. *Journal of Clinical Investigation*, **14**, 266–75.

EPSTEIN, A. N., and P. TEITELBAUM (1964). Severe and persistent deficits in thirst produced by lateral hypothalamic damage. In *Thirst*, M. J. Wayner (ed.), 395–410. Pergamon Press.

EVANS, C. L. (1949). *Principles of Human Physiology*, Churchill.

FALK, J. L. (1961). The behavioural regulation of water electrolyte balance. *Nebraska Symposium on Motivation*, **9**, 1–33, University of Nebraska Press.

FINCHAM, E. F., and J. WALTON (1957). The reciprocal actions of accommodation and convergence. *Journal of Physiology* (*London*), **137**, 488–508.

FITTS, P. M., and M. I. POSNER (1967). *Human Performance*, Brooks/Cole Publishing Co., California.

FITZSIMONS, J. T. (1961). Drinking by rats depleted of body fluid without increase in osmotic pressure. *Journal of Physiology* (*London*), **159**, 297–309.

FITZSIMONS, J. T., and K. OATLEY (1968). Additivity of stimuli for drinking in rats. *Journal of Comparative and Physiological Psychology*, **55**, 145–54.

GAUER, O. H., and J. P. HENRY (1963). Circulatory basis of fluid volume control. *Physiological Reviews*, **43**, 423–81.

GILMAN, A. (1937). The relation between blood osmotic pressure, fluid distribution and voluntary water intake. *American Journal of Physiology*, **120**, 323–8.

GORDON, M. S. (1972). *Animal Physiology: Principles and Adaptations*, Macmillan.

GREGORY, R. (1966). *Eye and Brain*, Weidenfeld and Nicolson.

GUYTON, A. C. (1969). *Function of the Human Body*, Saunders, Philadelphia.

GUYTON, A. C. (1971). *Textbook of Medical Physiology*, Saunders, Philadelphia.

GUYTON, A. C., and T. G. COLEMAN (1967). Long term regulation of the circulation interrelationships with body fluid volumes. In *Physical Bases of Circulatory Transport: Regulation and Exchange*. E. B. Reeve and A. C. Guyton (eds.), Saunders, Philadelphia.

GUYTON, A. C., and W. M. GILLESPIE (1951). Constant infusion of epinephrine: rate of epinephrine secretion and distribution in the body. *American Journal of Physiology*, **165**, 319–27.

HATTON, G. I., and C. T. BENNETT (1970). Satiation of thirst and termination of drinking: roles of plasma osmolality and absorption. *Physiology and Behaviour*, **5**, 479–87.

HERSTEIN, I. N., and R. SANDLER (1971). *Introduction to the Calculus*, Harper and Row, New York.

HOLMES, J. H., and GREGERSEN, M. I. (1950). Role of sodium and chloride in thirst. *American Journal of Physiology*, **162**, 338–47.

HOLST, E. von (1954). Relation between C.N.S. and the peripheral organs. *Animal Behaviour*, **2**, 89–94.

HSAIO, S. (1970). Reciprocal and additive effects of hyperoncotic and hypertonic treatments on feeding and drinking in rats. *Psychonomic Science*, **19**, 303–4.

HULL, C. L. (1943). *Principles of Behaviour*, Appleton-Century, New York.

KLOOT, W. G. van der (1968). *Behaviour*. Holt, Rinehart and Winston, New York.

LANGE, G. W. (1967). Some characteristics of human operator tracking. *I.E.E. Colloquium on Methods of Interpreting Biological Signals*, 25–31.

LOWENSTEIN, O., and I. E. LOEWENFELD (1953). Effect of physostigmine and pilocarpine on iris sphincter of normal man. *Archives of Ophthalmology*, **50**, 311–18.

MCCANCE, R. A. (1936). Experimental sodium chloride deficiency in man. *Proceedings of the Royal Society, London*, **B119**, 245–68.

MCCORMICK, E. J. (1957). *Human Factors Engineering*, McGraw-Hill, New York.

McFarland, D. J. (1971). *Feedback Mechanisms in Animal Behaviour*, Academic Press.

McFarland, D. J., and P. Budgell (1970). Determination of a behavioural transfer function by frequency analysis. *Nature*, **226** 966–7.

McFarland, D. J., and F. J. McFarland (1968). Dynamic analysis of an avian drinking response. *Medical and Biological Engineering*, **6**, 659–67.

MacKay, D. M. (1956). Towards an information flow model of human behaviour. *The British Journal of Psychology*, **47**, 30–43.

Milhorn, H. T. (1966). *The Application of Control Theory to Physiological Systems*, Saunders, Philadelphia.

Miller, N. E., R. I. Sampliner, and P. Woodrow (1957). Thirst reducing effects of water by stomach fistula vs. water by mouth measured by both consummatory and an instrumental response. *Journal of Comparative and Physiological Psychology*, **50**, 1–5.

Milsum, J. H. (1966). *Biological Control Systems Analysis*. McGraw-Hill, New York.

Mittelstaedt, H. (1957). Prey capture in mantids. In *Recent Advance in Invertebrate Physiology*, B. T. Scheer (ed.), University of Oregon.

Morgan, M. W. (1968). Accommodation and convergence. *American Journal of Optometry*, **45**, 415–54.

Nakayama, T., H. T. Hammel, J. D. Hardy and J. S. Eisenman (1963). Thermal stimulation of electrical activity of single units of the preoptic region. *American Journal of Physiology*, **204**, 1122–6.

Neisser, U. (1966). *Cognitive Psychology*. Appleton-Century, New York.

Novin, D. (1962). The relationship between electrical conductivity of brain tissue and thirst in the rat. *Journal of Comparative and Physiological Psychology*, **55**, 145–54.

Oatley, K. (1964). Changes in blood volume and osmotic pressure in the production of thirst. *Nature*, **202**, 1341–2.

Oatley, K. (1965). Thirst and drinking mechanisms in the regulation of water intake in rats. Ph.D. thesis, University of London.

Oatley, K. (1967). A control model of the physiological basis of thirst. *Medical and Biological Engineering*, **5**, 225–37.

OATLEY, K. (1972). Book review of D. J. McFarland, 'Feedback Mechanisms in Animal Behaviour', *Quarterly Journal of Experimental Psychology*, **24**, 553–4.

OATLEY, K. (1973). Simulation and theory of thirst. In *The Neuropsychology of Thirst*, A. N. Epstein, H. R. Kissileff, and E. Stellar (eds), Winston.

OATLEY, K. (1974). Circadian rhythms in models of motivational systems. In *Methods of Motivational Control Systems Analysis*, D. J. McFarland (ed.), Academic Press.

OATLEY, K., and A. DICKINSON (1970). Air drinking and the measurement of thirst. *Animal Behaviour*, **18**, 259–65.

OATLEY, K. and F. M. TOATES (1969). The passage of food through the gut of rats and its uptake of fluid. *Psychonomic Science*, **16**, 225–6.

OATLEY, K., and F. M. TOATES (1971). Frequency analysis of the thirst control system. *Nature*, **232**, 562–4.

OATLEY, K., and F. M. TOATES (1973). Osmotic inhibition of eating as a subtractive process. *Journal of Comparative and Physiological Psychology*, **82**, 268–77.

O'KELLY, L. I., J. L. FALK, and D. FLINT (1958). Water regulation in the rat: 1. Gastrointestinal exchange rates of water and sodium chloride in thirsty animals. *Journal of Comparative and Physiological Psychology*, **51**, 16–21.

PANKSEPP, J., F. M. TOATES and K. OATLEY (1972). Extinction induced drinking in hungry rats. *Animal Behaviour*, **20**, 493–9.

PRIBAN, I. P., and W. F. FINCHAM (1965). Self-adaptive control and the respiratory system. *Nature*, **208**, 339–43.

RASHBASS, C., and G. WESTHEIMER (1961). Disjunctive eye movements. *Journal of Physiology* (*London*), **159**, 339–60.

REEVE, E. B., and L. KULHANEK (1967). Regulation of body water content: a preliminary analysis. In *Physical Bases of Circulatory Transport: Regulation and Exchange*, E. B. Reeve and A. C. Guyton (eds.), 157–77. Saunders.

RUCH, T. C. (1951). Motor Systems. In *Handbook of Experimental Psychology*, S. S. Stevens (ed.), 154–208. Wiley, New York.

SMITH, H. W. (1957). Salt and water volume receptors. *American Journal of Medicine*, **23**, 623–52.

SPERRY, R. W. (1951). Mechanisms of neural maturation. In *Handbook of Experimental Psychology*, S. S. Stevens (ed.), Wiley, New York.

STARK, L., and P. SHERMAN (1957). A servo-analytical study of consensual pupil reflex to light. *Journal of Neurophysiology*, **20**, 17–26.

STARK, L., Y. TAKAHASHI and G. ZAMES. (1965). Non-linear servo-analysis of human lens accommodation. *I.E.E.E. Transactions on Systems Science and Cybernetics*, **SSC-1**, 75–83.

STELLAR, E. and J. H. HILL (1953). The rat's rate of drinking as a function of water deprivation. *Journal of Comparative and Physiological Psychology*, **45**, 96–102.

STRICKER, E. M. (1966). Extracellular fluid volume and thirst. *American Journal of Physiology*, **221**, 232–8.

STRICKER, E. M., and G. WOLF (1969). Behavioural control of intravascular fluid volume: thirst and sodium appetite. *Annals of the New York Academy of Sciences*, **157**, 553–68.

SUTHERLAND, N. S. (1972). Book review of J. W. Kling and L. A. Riggs, 'Woodworth and Schlosberg's Experimental Psychology', *Nature*, **240**, 167–8.

SWAN, K. C., and L. GEHRSITZ (1951). Competitive action of miotics on the iris sphincter. *Archives of Ophthalmology*, **46**, 477–481.

TOATES, F. M. (1970). A model of accommodation. *Vision Research*, **10**, 1069–76.

TOATES, F. M. (1971). Thirst and body fluid regulation in the rat. D.Phil. thesis, University of Sussex.

TOATES, F. M. (1972a). Accommodation function of the human eye. *Physiological Reviews*, **52**, 828–63.

TOATES, F. M. (1972b). A model of an autonomic effector control loop. *Measurement and Control*, **5**, 354–7.

TOATES, F. M. (1972c). Further studies on accommodation and convergence. *Measurement and Control*, **5**, 58–61.

TOATES, F. M. (1974a). Vergence eye movements. *Documenta Ophthalmologica* **37**, 153–214.

TOATES, F. M. (1974b). Computer simulation and the homeostatic control of behaviour. In *Methods of Motivational Control Systems Analysis*, D. J. McFarland (ed.), Academic Press.

TOATES, F. M., and K. OATLEY (1970). Computer simulation of thirst and water balance. *Medical and Biological Engineering*, **8**, 71–87.

TOATES, F. M., and K. OATLEY (1972). Inhibition of ad libitum feeding in rats by salt injections and water deprivation. *Quarterly Journal of Experimental Psychology*, **24**, 215–24.

WELFORD, A. T. (1968). *Fundamentals of Skill*. Methuen.

WESTHEIMER, G. (1954). Mechanism of saccadic eye movements, *Archives of Ophthalmology (NY)*, **52**, 710–24.

WIENER, N. (1948). *Cybernetics*. MIT Press, Mass.

WILKINS, B. R. (1966). Regulation and control in engineering. In: *Regulation and Control in Living Systems*, H. Kalmus (ed.), 12–28. Wiley.

WOLF, A. V. (1958). *Thirst: Physiology of the urge to drink and problems of water lack*. Thomas, Springfield.

YOUNG, L. R., and L. STARK (1963). Variable feedback experiments testing a sampled data model for eye tracking movements. *I.E.E.E. Transactions on Human Factors in Electronics*, **4**, 38–51.

ZUBER, B. L., and L. STARK (1968). Dynamical characteristics of the fusional vergence eye-movement system. *I.E.E.E. Transactions on Systems Science and Cybernetics*, **SSC-4**, 72–9.

INDEX